D0227926

Horse and Stable
Management

Also from BSP Professional Books

Getting Horses Fit
Sarah Pilliner
0 632 02307 4

Equine Injury and Therapy
Mary Bromiley
0 632 02277 9

Keeping Horses
How to save time and money
Susan McBane
0 632 02363 5

Practical Stud Management
John Rose and Sarah Pilliner
0 632 02031 8

How to Keep Your Horse Healthy
Colin Vogel
0 632 02056 3

Horse and Stable Management

Jeremy Houghton Brown
and
Vincent Powell-Smith

BSP PROFESSIONAL BOOKS
OXFORD LONDON EDINBURGH
BOSTON PALO ALTO MELBOURNE

Copyright © Jeremy Houghton Brown
and Vincent Powell-Smith 1984

All rights reserved. No part of this
publication may be reproduced, stored
in a retrieval system, or transmitted, in
any form or by any means, electronic,
mechanical, photocopying, recording or
otherwise without the prior permission
of the copyright owner.

First published in Great Britain
by Granada Publishing 1984
(ISBN 0–246–11217–4)
Reprinted 1985 with minor amendments
by Collins Professional Books
Reprinted 1986, 1987 (twice)
Reprinted by BSP Professional Books 1987,
1988 (twice)

Illustrations by Jeremy Houghton
Brown

British Library Cataloguing in
Publication Data

Houghton Brown, Jeremy
  Horse and stable management.
  1. Horses
  I. Title   II. Powell-Smith, Vincent
  636.1'083   SF285.5

ISBN 0–632–02141–1

BSP Professional Books
Editorial offices:
Osney Mead, Oxford OX2 0EL
   (Orders: Tel. 0865 240201)
8 John Street, London WC1N 2ES
23 Ainslie Place, Edinburgh EH3 6AJ
52 Beacon Street, Boston
   Massachusetts 02108, USA
667 Lytton Avenue, Palo Alto
   California, 94301, USA ·
107 Barry Street, Carlton,
   Victoria 3053, Australia

Set by V & M Graphics, Aylesbury,
Bucks.

Printed in Great Britain by
Hollen Street Press Ltd, Slough

To JANE, who proves that theory and good practice are complementary by training horses to achieve success

J. H. B.

To LIZ, who has long put into practice the principles of good management, and to KORBUT, who masters the Moor.

V. P.-S.

Contents

Acknowledgments

The authors wish to acknowledge the help and assistance given to them by their respective wives, as well as that of colleagues and friends. Particular thanks are owed to Gillian McCarthy BSc (Hons) and Angela Boyden, and to Stewart Hastie MRCVS, who has done so much to encourage the concept that the best horse care comes from understanding both the body and the mind of the horse.

Foreword

In asking me to write a foreword for this book the authors have
paid me a great compliment. I read it with interest and I consider
it a book which has been needed for a long time. The contents are
subjects which every caring horse owner or horsemaster should
know in detail. The authors have written this book in a very clear
way which can be easily understood and used quickly as a book of
reference for most equestrian problems.

The chapters on the Healthy and Sick Horse are extremely
comprehensive as well as the chapters on the Systems of the Horse
which are usually very difficult to find in any one book. The
chapters on The horse at grass and The horse under stress discuss
subjects I have never seen written down in such a sensible way
before. We are putting much greater demands on the horse in this
modern world and competitions are asking more and more athletic
ability from the horse, so better breeding, correct handling and
greater knowledge are required from us all as it is our duty to try
to understand the sort of stresses and strains we ask of the horse in
this highly competitive world. Jeremy Houghton Brown and
Vincent Powell-Smith have considerable knowledge and experience in all of these fields having made a study of the horse over
many years.

'Horse and Stable Management' is especially useful for the
serious student and anyone considering any horse examinations.
Many people go forward for examinations sadly lacking in
sufficient knowledge of horse management and the functions of the
horse, and this book will be of great help to such people.

One would find it very difficult to obtain all this information
anywhere and to have it available in this book is extremely useful.
Anyone considering a career with horses should certainly think of
this book as an invaluable part of their collection.

Jennie Loriston-Clarke MBE,FBHS

Preface

This book is intended for people who own or work with horses or who wish to know more about the horse. It explains how the horse functions as well as discussing horse and stable management. It is not a first horse book, because we have assumed a basic standard of knowledge and experience. For this reason, it will aid the student as well as others who care for horses. Indeed, this book will also serve as a textbook for examinations set by the British Horse Society, the National Pony Society, and the Association of British Riding Schools, and for National Certificates and Diplomas.

Good horse management, sometimes called horsemastership, means efficiently and pleasantly getting the best out of the horse or pony in all seasons and on all occasions, be it in the stable, at grass, in competition, at stud, in sickness, in health, in youth or in maturity.

Although this book touches on the fitness and exercise of the horse, it does not cover riding, and thus we have concentrated on the 'stable management' aspects in some detail in order to give greater understanding.

The horse is a complex of systems, all of which are discussed here. A problem in one system can have repercussions in others, and cause complications. If a sick horse is to be made well, the first and essential step is to identify the cause of the problem. Careful examination of the horse will reveal some of the symptoms but it is pointless to treat them and leave the cause. Only by studying the systems when they are working well, and understanding how each one works, can one really appreciate what goes wrong and – more importantly – when it is likely to do so.

The good horsemaster, man or woman, understands and cares for the horse in such a way that all the systems co-ordinate and function efficiently. In this way, the horse is enabled to give of its best.

In the text, there is a mixture of horseman's jargon and veterinary terminology, as well as discussion of some scientific concepts, but the accent is always on sound, practical care. A further aim is to help with the vital decision of when to call the vet, a term (generally of endearment) for a veterinary surgeon, and to give a better understanding of what he or she will do, and of how to provide the best back-up to this service. However, nothing in this book should encourage the reader to try to become a 'do-it-yourself' vet; indeed, our main aim is to promote management with understanding.

Jeremy Houghton Brown
Vincent Powell-Smith

Part I

The Horse in Sickness and Health

1 The healthy horse

What is Health?

In its natural state a horse strives to do three fundamental things: to survive, to nourish itself and to reproduce. When frightened it flees or, if cornered, it fights. It eats and drinks in response to hunger and thirst and thus grows to maturity and maintains its strength. In response to sexual desire it endeavours to reproduce and so continue its species. These three things are key points which are characteristic of a healthy horse. It has a good appetite and a digestive system that works well. It grows strong and fit for the work to which it is accustomed. It is alert and perceptive. If not interfered with, it is eager and able to breed. Thus, health is not merely freedom from disease; it is also a state of well-being and vigour. In practical terms, the healthy horse must be able to give of its best consistently, whether in terms of producing a foal yearly, of running races well, or of hunting three days a fortnight through the Season.

The key to maintaining health is to be observant and perceptive. Anyone who looks after animals must learn to develop that great gift, 'the stockman's eye'. This is the ability to note the normal look, feel and behaviour of an animal so that any difference is spotted at once. To keep a horse healthy it is vital to know how it looks and behaves normally. This is particularly important when the horse is undergoing change, as is the case with pregnant mares, foals, growing stock, horses just up from grass, horses being roughed off and those being fittened.

Signs of Health

Behaviour

A group of horses generally acts as a herd. An isolated horse or one uninterested in the behaviour of its fellows is abnormal. Mature horses in fields generally remain standing during the day and often have favourite spots for dozing. It is usual for horses to doze for about a third of the time. Each nap is generally of short duration. When they lie down, which happens more frequently in stables than in the open, horses go down front end first with bent front legs. They get up front end first with straight front legs. Horses rest with the breast bone to one side and can only change sides by rolling their legs up and over or by getting up and lying down again.

Appetite

When food is available continuously, the horse prefers to eat little and often. It is therefore normal for the horse at grass to graze intermittently day and night. The stabled horse normally goes straight to its short feed and eats it all. Some stabled horses are shy feeders and eat slowly when left in peace; others are anxious and eat best if the manger is hung on the door. After resting, foals normally get up and go straight to their mother for a drink. The vast majority of horses relish good food: any horse that does not must arouse suspicion about its health. Horses' drinking habits can also indicate their health, and because of this it is best not to use self-filling drinkers for off-colour horses.

Dung and Urine

Dung should be green–brown to golden brown depending on the diet, it should be moist, and it should break slightly on hitting the ground. Any changes from usual quantity, colour or pattern of behaviour should be noted.

Eyes and Ears

The eye should be bright and wide open and the ears mobile. These organs give good indications of alertness and interest. Any failure to respond to sights or sounds may be an indication that something is wrong.

Body Processes

Temperature, pulse and respiratory rate (T.P.R.) are not obvious in a healthy horse at rest. If any one of these is unusually apparent, then the cause must be sought. When a horse is at rest the pulse rate, being a reflection of heart rate, will be at its slowest for that horse – nature's economy at work. The normal rates for an adult horse at rest are: temperature 38°C (100.5°F); pulse about 40 beats per minute; respiration about 12 breaths per minute.

The horse's temperature is taken in the rectum with a clinical thermometer. The procedure is as follows:

(1) Shake the thermometer so that it reads several degrees lower than normal.

(2) Lubricate the bulb with vaseline or saliva. Stand behind but to one side of the horse to avoid being kicked.

(3) Hold the horse's tail to one side and insert the bulb of the thermometer gently into the horse's anus, rotating it slightly as you do so. It should be inserted to half-way, at a slight angle to press the bulb against the side of the rectum.

(4) Leave in position for 1 minute.

(5) Withdraw and read the thermometer, taking care not to hold it at the bulb end. The mercury is most clearly seen if viewed through the apex which runs the full length of the thermometer.

(6) Clean and disinfect the thermometer before returning it to its case.

The temperature may safely vary half a degree from normal, and in foals the norm can be as high as 38.6°C (101.5°F).

The pulse can be taken by pressing the fingers against an artery passing over a bone close to the surface, e.g. the facial artery on the inside edge of the lower jaw (see Fig. 1.1) or on the radial artery inside the foreleg, level with the elbow. Some people like to take the pulse under the dock while taking the temperature. The simplest method of taking the pulse is to use a stethoscope just behind the horse's left elbow; however, even a flat hand lightly placed there may feel the heart beat. The importance of monitoring pulse in such sports as long-distance riding has brought the use of the stethoscope into the range of the layman. The pulse rate of individual horses varies: a rate of between 35 to 45 a minute may be normal for a particular horse at rest, and a rate of 50 to 100 a minute for a foal.

Positions for feeling
the pulse with the
finger tips resting
lightly on the artery

Artery crossing the jaw bone

Inside of foreleg just in front of elbow

Fig. 1.1 Positions for feeling the pulse with the fingertips resting lightly on the artery.

To observe respiration, the horse must be standing still and undisturbed. Watch the horse's flanks from the side. Each complete rise and fall is one breath. On a chilly day, respiration can be observed at the nostrils as condensation in the air. A range of 8 to 16 breaths a minute is acceptable in the adult horse at rest; 20 to 30 breaths a minute is acceptable in the foal. It is important to monitor the temperature, pulse and respiration (T.P.R.) of a new arrival in the stable, doing so at the same time each day (preferably first thing in the morning) for several days, so that one knows what is normal for that individual.

Maintaining Health

Domestication has deprived horses of some of their natural means of staying healthy. The wandering, grazing herd gets steady exercise, intake of a widely mixed diet little and often, ready access to water, and freedom from mental and physical stress. Some aspects of maintaining health merit individual consideration at this stage. However, feeding is so complex and important that it is deferred until later: see chapter 10 and Appendix.

Air

Hot, stuffy, stale or limited air all predispose to respiratory problems. Horses must have plenty of fresh air at all times, but they dislike draughts. They do not mind the air being cold, providing it is relatively still. In racing stables, the practice of shutting top doors at certain times is to ensure peace and quiet to each horse; each of these stables has independent air access, and even those stables warmed by infra-red lamps have access for plenty of fresh air.

Exercise and Rest

In the wild the horse takes exercise gently but steadily throughout the day. In a field, the horse is able to wander as it grazes and this exercise is sufficient to aid digestion and circulation. Ideally, the stabled horse should have a period in the field every day, but in larger establishments this would pose too many problems and risks. The alternative is daily exercise, ridden or in hand. When horses spend even one day a week in the box, this can lead to the so-called 'Monday morning diseases' such as azoturia and lymphangitis.

When hard work is in progress, blood is in great demand in the muscles and is therefore drawn away from the digestive system. Fast work also calls for greater use of the diaphragm to assist in the maximum intake of air. Since the stomach presses against the diaphragm, it is important that it is relatively empty when the horse is working. It is therefore usual not to do work for at least an hour after feeding.

Exercise must be built up slowly so that the horse can cope efficiently with the demands without 'running up light' or getting 'tucked up'. (These terms are explained later in the chapter.) Each horse has its own physical limits and special talents. It is up to the owner to get the greatest benefit from each individual animal.

The horse must get proper rest as well as exercise. This principle is well understood in racing yards, with their afternoon 'siesta' period of quiet in the yard. Stabled horses generally enjoy a rest during the day. A comfortable bed and a peaceful and orderly yard will encourage this. The natural pattern of about three hours' grazing, followed by a rest of up to two hours, is the ideal for good digestion and food utilisation.

Protection

The wild horse has natural grease in its coat to protect it from the weather, and it can seek shelter from the prevailing wind and from flies. In a field, the horse either has the grease left in its coat or is provided with at least a New Zealand rug to keep out the wet. It may also be provided with a field shelter. The stabled horse is not free to wander, and if there is a draught in the stable the horse cannot escape it. Inadequate bedding will cause the horse to bruise itself and get cold and damp when it lies down. A stabled horse that feels chilly or stiff cannot take exercise to stimulate the circulation, ease the joints and keep itself warm. If its circulation is sluggish and its legs start to fill, the stabled horse relies on its keeper for its exercise. The horse owner must accept full responsibility for his or her animal's exercise, warmth and protection.

Peace of Mind

In order to maintain health in the horse, its peace of mind must be considered. This subject is not yet well documented or researched, but common sense and consideration are the keynote.

The horse is a herd animal which seeks refuge in flight. The herd has natural leaders who insist on discipline and obedience, and within the herd the mares will require these things of their foals. By observing horses, the owner can learn how best to treat them.

The horse tends to do better with company, regular routine, fair punishment when it does wrong, and reward when it does right, with consistency being the most important factor.

Conformation

Conformation is the horse's inherited structure. It is the way the horse is put together. Beauty is said to be 'in the eye of the beholder' and some people watching the judging of a show class may wonder if conformation is equally difficult to define. Conformation is certainly a matter of opinion, but if we agree on what is meant by conformation, then the horseman can build up mental pictures and a list of criteria.

Conformation has two main aspects. The first is the shape of the horse and is called 'static conformation'. The second aspect is the way the horse moves, and is called 'dynamic conformation'. The horse's performance, in terms of speed, endurance, jumping ability, agility, and so on, is not really included in conformation. Temperament factors, such as obedience, kindness and generosity, are taken into slight consideration when assessing dynamic conformation. The important factor is that the horse should conform to a pattern proved over the years to produce the best performance, and not be predisposed to weakness, illness or disease. Consideration of conformation is thus an essential part of the attention given to maintaining a healthy horse.

'Good conformation' in various types of horse can be seen in the show ring. First impressions are very important and, although turn-out can help, the discerning eye will concentrate on the shape of the skeleton and of the muscle clothing it.

Static Conformation

The Head

The head must be in proportion to the size of the horse. If it is too big, the horse will always tend to be on its forehand. The rounded convex profile of a Roman nose may indicate common blood and a dished profile may be a sign of Arab blood. The lower and upper jaws should meet evenly at the front, so the lips should be drawn back to check. If the upper jaw is too long, the horse is said to be 'parrot mouthed', and this is an unsoundness. Conversely it may be 'sow mouthed', when the lower jaw is too long. The terms 'overshot' and 'undershot' are confused in some textbooks and so are best avoided. In either case the fault will affect the horse's ability to bite food such as grass, but not its ability to chew.

Experience suggests that a bold eye indicates a generous spirit and a small 'piggy' eye may indicate meanness. The eyes should be set well out at the side of the head, and be clear, large and prominent. An excess of white in the eye suggests the possibility of an excitable and ill-tempered nature. Lop ears are not a fault, although such horses may need more encouragement.

Next to size, the most important aspect of the head is the way it is set on to the neck. There must be adequate clearance between the wing of the top bone in the neck and the branch of the lower jaw. This clearance should be sufficient to take two fingers when the horse's head is raised, and it must not restrict the flexion at the poll when it is brought into a position where the front of the face runs vertically, as in the collected gaits. The horse must not be fleshy around the jowl. The muzzle should be fine, with thin lips and sound and regular incisor teeth. The nostrils should be large.

The Neck

A long neck goes with elegance, and a shorter, heavily muscled one with strength. From the rider's point of view it is not comfortable to have 'the horse's head in one's lap', and the judge of a ridden horse assesses the horse's 'front' as much from the saddle as from the side. If the neck dips down in front of the withers, the horse is 'ewe-necked', and this makes it more difficult to achieve the desirable forward, steady, head carriage, with flexion at the poll and a relaxed lower jaw. There should be an unbroken curve from the poll to the withers.

The Shoulder

The shoulder starts at the withers with the cartilage extension of the shoulder blade or scapula, which runs forward to the point of the shoulder. (See Fig. 1.2.) The line from the withers to the point of the shoulder is known as the slope of the shoulder. An upright shoulder gives a short stride and the front legs will show wear more quickly. A good sloping shoulder is necessary. From the point of the shoulder the upper arm bone or humerus runs down and back to the elbow. In the riding horse the whole shoulder should be well muscled and yet without heaviness.

The Front Legs

The forearm should be well muscled from the elbow, and length is required for speed. The knee is a box of bones, each separated by shock-absorbing cartilage. Knees should be broad and flat in front. The fault of 'calf knees' means they are shallow from front to back. The leg should not appear to be 'tied in' as if the horse were wearing tight stockings restricting the tendons below the knee.

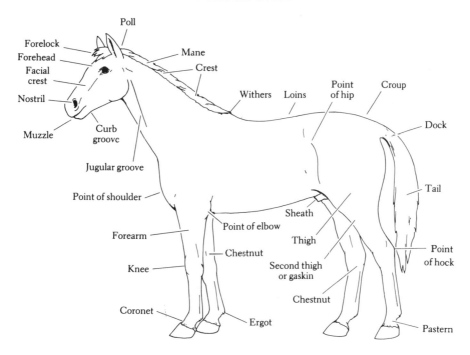

Fig. 1.2 Points of the horse.

The measurement around the leg just below the knee is used to define the amount of 'bone' the animal has. This measurement includes the tendons. Common-bred horses will have more bone. The normal aim is to combine quality with good bone. As a rough guide, in a horse of 16.2 hands there should be over 20 cm (8 in) of bone for a lightweight and over 23 cm (9 in) for a heavyweight.

When viewed from the front and side, the knee joint should be straight. Viewed from the side, if the knee is slightly forward of the line from the elbow to the top of the pastern, the horse is 'over at the knee'. This may not please the show judge, but it will increase the shock-absorbing properties of the front legs. The reverse case of 'back at the knee' is to be avoided as the strain on the tendons may prove too great.

The cannon bones should be rather short and flat in front, with the tendons standing out cleanly at the back. This ensures that the tendons are short and less liable to damage. The slope of the pastern is important: too much slope puts too great a strain on the tendons, and too little creates excess concussion and leads to foot troubles.

The feet must be good. 'No foot no horse' is a true adage. Upright feet are called 'boxy' and are to be avoided. Flat feet are equally undesirable. The angle of the hoof wall should be such that it continues the line of the pastern. There should be a good frog on the underside of the foot, and the bars and the heel of the hoof must be wide and deep. Any irregularity of the feet may cause or be the result of foot trouble. The front legs should be checked from the front to see if they are upright and straight from the body right into the feet. The horse must not be knock-kneed or bandy. The feet should face the front and not be turned in ('pigeon toed') or turned out ('splay footed').

The Chest and Barrel

On a riding horse the withers should be clearly defined. High withers may make it difficult to fit the saddle. A deep, full chest with long, 'well-sprung' ribs is essential to provide good lung- and heart-room. A horse with flat ribs is 'slab sided'. A horse of 16 hands or more should have a girth that exceeds 1.83 m (6 ft). The front legs must have clear space between them and not 'come out of one hole'. On the other hand, if the chest is too wide, it may produce a rolling action.

The back may be dipped or hollow from old age or from lack of good conformation. A 'roach back' is one curved upwards and might make saddle-fitting difficult. The back should also be checked with a saddle to see whether the horse sinks under the rider's weight when first mounted or even when girthing up. Such a horse is said to be 'cold backed'.

Good lung- and heart-room is the first essential of the body, which must not be weak over the loins. The first eight pairs of ribs connect up to the breast bone and are called 'true ribs'. The next ten pairs of ribs are connected by long cartilage extensions to the breast bone and they are known as 'false ribs'. Some horses have a nineteenth rib on one or both sides. The rearmost ribs must come close to the point of hip so that the horse is 'well ribbed up'. A wide distance between the last rib and the point of hip makes a horse 'slack in the loins', and such a horse is sometimes described as 'short of a rib'.

It is a weakness when the underline of the horse slopes up from front to back (described as being 'herring gutted'). This fault may cause the girth to slip back. The underline of the horse may also vary according to the diet and degree of fitness. A horse at grass normally has a very full gut and may even be 'pot-bellied'. After stress, there may be a tightening of the tummy muscles for several hours and the horse is accurately described as 'tucked up'. This is not the same as 'run up light', which denotes the overall leanness which may go with hard training, hard work or lack of food.

The top line of the horse should be such that, once the horse is fully grown, the withers are higher than the croup. In a young horse the croup is often higher, but if this persists into maturity, the horse will ride 'down hill' and tend to be on its forehand.

The Hindquarters

The powerhouse of the horse lies behind the saddle and so one looks on the quarters for good flat muscle which reaches well down into the hind leg. A high croup is called a 'jumping bump'. A 'goose rump' is one with the tail set low down and the quarters drooping downwards. Horses with 'goose rumps' are thought to lack speed. Good speed is expected from horses with plenty of length from the point of hip to the hocks; thus one may refer to 'well-let-down hocks'. The horse must not appear split up the middle when viewed from behind; the thigh muscles of the hind legs must be well developed on the inside of these limbs. The human knee joint is the equivalent of the horse's stifle joint: both have a patella or kneecap. With a human runner we note good muscle development above and below the knee; so with the horse we look for good development of the gaskin or second thigh muscles, which run from the stifle to the hock.

Hocks must not point towards each other (a fault described as being 'cow hocked') nor must they be 'bowed out'. They must not be over bent so that the cannon bones slant ('sickle hocked'). The hock joint should be large but not fleshy. The line from the point of hock down the back of the leg should be vertical and should not bulge outwards over the hock joint. A bulge of this sort is called a 'curb' and is usually caused by strain on the tendon. The bulge could also be caused by enlarged heads of the splint bones and is then called a 'false curb'. Curbs are considered a sign of weakness, but generally give little trouble once they have formed on the young horse. The vertical line below the hock should, when the horse is standing squarely, line up with the rearmost part of its quarters (point of the buttock). The comments concerning the

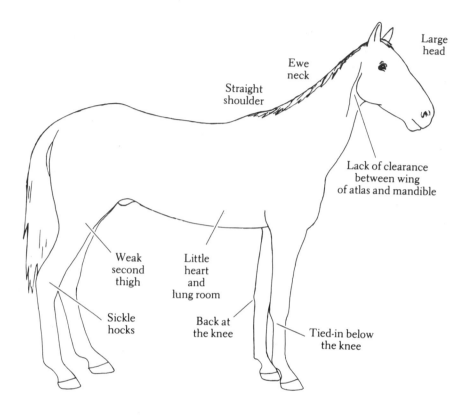

Fig. 1.3 Poor conformation.

lower forelimb apply equally to the lower hindlimb.

Poor conformation is illustrated in Fig. 1.3.

Dynamic Conformation

The walk and trot should be checked both in hand and under saddle.

The Walk

The walk is in four time. Regular steps of even length are required. The footprint of the front foot should be studied to see whether the back foot comes on to it (tracking up) or, even better, goes beyond it (overtracking). The amount by which it overtracks should be noted. A good walker will give the impression that it is going somewhere in a purposeful manner. The walk is a difficult pace to improve. A good walker is generally a good galloper. The way the shoes are worn gives clues as to how the horse uses its feet.

The Trot

The trot is in two time, and when there is any extension there is a moment of suspension between each beat. The horse should be trotted towards the observer, then on past and away from him or her. The action of the front legs should be noted. A straight action is best. To swing the feet out from the knee or fetlock joint is called 'dishing' and is unsightly and energy wasting although it is not harmful. Any action that brings one foot up too close to the other leg, which could result in 'brushing' or, if higher, in 'speedy cutting', is to be avoided. The hind feet often pass close by each other and one can check the hair on the inside of the hind fetlock joints. If it is rubbed, the horse may require Yorkshire boots or other protection. When the horse trots, any tendency to 'forge' should be listened for. Forging is the hind shoe striking the fore shoe. Both forging and over-reaching can be helped by shoeing. Viewed from the side, the strides of the left diagonal, near fore and off hind, should be the same length as the right diagonal. The amount of knee action required depends on fashion and use. However, where knee action is required (as with some driving and draught horses), the animal must still cover the ground well. The show horse tends to push its foot out a long way, and then drops it vertically down the last few inches. A good riding horse will push its toe out well without any exaggeration of movement. It will not land heavily on its heels. A free moving shoulder with no sign of cramped or restricted movement when the horse is asked to lengthen the trot is desirable. A supple back and well-engaged hindquarters, with freely flexing hocks, are needed.

The Canter

This gait is in three time, followed by a moment of suspension. Balance comes from the hind legs coming well under the horse. The young or newly broken horse may find it difficult to canter in a confined area. At the canter one can also listen for soundness of the wind.

The Gallop

Some horses achieve the rhythm of the gallop but seem to be 'going into the ground'. Look for the horse that achieves both speed and lightness.

The Step Back

When checking soundness, the horse is asked by its handler to step back on the ground and also to turn in tight circles to left and right. These movements may reveal any stiffness or other problems. Conformation and soundness are relative to each other.

Legal Unsoundness

Unsoundness is a question of usefulness and not of disease. The position in English law was stated in the 19th century case of *Coates* v. *Stephens* in this way:

> The rule as to unsoundness is, if at the time of sale or examination, the horse has any disease, which either actually does diminish the natural usefulness of the animal, so as to make him less capable of work of any description, or which in its ordinary course will diminish its natural usefulness; or if that horse has, either from disease (whether such disease be congenital, or arising subsequently to birth) or from accident has undergone any alteration of structure that either does at the time or in its ordinary course will diminish the natural usefulness of the horse, such horse is unsound.

Despite the antique flavour of the judicial language, this is still the legal position today.

The Royal College of Veterinary Surgeons and the British Veterinary Association try to discourage their members from using the word 'sound'. This is because the veterinary profession had become increasingly anxious over the years at the way in which Baron Parkes' famous definition of soundness was being interpreted by the courts. It is one thing to ascertain if a horse has, at the present time, a disease or a defect which diminishes its natural usefulness. It is quite another to be certain whether it has some latent or minor defect of disease condition which, in its ordinary progress, will diminish its natural usefulness in the future.

Moreover, with all the sophisticated diagnostic aids available today, it is an open question as to the lengths in effort, time and expense to which the vet must go in order to discover the presence of some defect or disease condition which is not ascertainable on an ordinary clinical examination.

As a result, vets now have a recommended form of examination of a horse for a purchaser, after which they are advised to conclude that the horse is or is not suitable for purchase for a particular use, e.g. as a child's pony, an eventer or whatever. The vet's certificate is given to the intending purchaser, and not to the vendor – and is related to the purchaser's specific intentions regarding the horse's use.

2 The sick horse

There are four main types of ill health. The first has *physical causes*, e.g. an injury resulting from an accident. The second type has *physiological causes*, e.g. the improper function of one part of the body. The third type of disorder arises from *nutritional causes*, which may be due to a deficiency in diet or to the ingestion of a poison, such as contaminated food or a poisonous plant. The fourth type of ill health is caused by an *invasion* of the horse's body by a living organism.

The Invaders

The principal invaders are shown in Fig. 2.1. The bacteria are described as pathogenic, i.e. causing disease in their host. There are millions of bacteria within the horse's gut which are beneficial and, indeed, the horse is totally dependent on them because they play an important role in digestion.

Germs

Invasions by bacteria and viruses (which collectively are called microbes or germs) display a similar pattern. Both are parasites by nature in that they live and multiply in the body, sometimes causing the death of their host. When these organisms enter the body, each cell divides and subdivides until there are millions of germs entrenched in the host. Only then do they make their presence felt and produce symptoms of disease. The time between infection and development of symptoms is called *the incubation period*: it may be a few hours or several months.

Some of these germs may find themselves in unsuitable circumstances; they then grow a thick protective coat and become spores. These are very hard to kill and can exist for years waiting

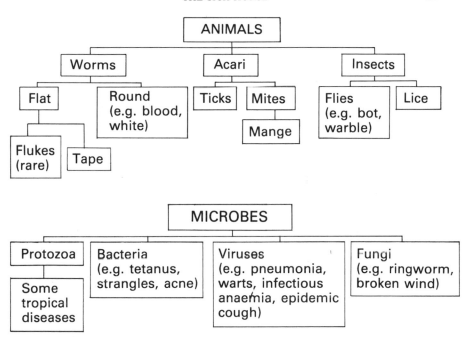

Fig. 2.1 Organisms that invade the horse.

for a suitable host. Once inside the host some germs invade the whole body; others stay localised in one organ. They vary in their ability to invade, persist and multiply, and this ability is called their *virulence*. Bacteria with a low virulence will cause a local infection such as an abscess, whereas one with a high virulence may invade the bloodstream and cause septicaemia (blood poisoning). Horses will be susceptible to certain germs but not to others; for example, horses cannot catch flu from humans.

Many bacteria and viruses produce poisons, called *toxins*, and these may spread through the body. When invaded, the host starts its defence programme against both the germs and the toxins. The animal is said to suffer an *acute attack* when its defences are inadequate and the disease is severe. If the defence is slow to react and merely holds the disease without overcoming it, the disease is *chronic*, i.e. longer lasting at a lower level.

The skin is the first line of defence. The second line of defence is the white blood corpuscles, which engulf bacteria and destroy them. This defence programme may cause inflammation and the production of pus, which is formed from the dead combatants. In addition to white corpuscles, the body can develop other defence agents, called *antibodies* which can combat only one type of germ. These are produced by the animal only when it encounters that particular germ. Further defence agents, called *antitoxins*, are produced to cope with the toxins produced by the invading germs. An animal is said to be *immune* to a disease when it has suffered it and has produced the specific defence agents to combat that disease. Such acquired immunity is of variable duration; in some cases it may be life-long.

Some disease-causing germs can be cultivated in a laboratory and then killed and let into the host, which acquires immunity once it has produced the necessary antibodies. The dead germs are prepared as a *vaccine*; being dead, they cannot multiply and thus they are easily overcome by the host, causing only minor symptoms.

For some diseases, dead germs will not achieve the desired effect, so the vaccine is prepared with live germs which have been weakened. In some cases, it may be necessary to repeat the process of vaccination initially after a short interval and then at regular, long-term intervals thereafter.

The vaccine may be swallowed, injected or inoculated by scratching the skin, according to type. For some diseases, modified toxins are used in vaccines. The host takes time to develop immunity after vaccination, and so if immediate protection is needed, blood serum will be injected. This will have been taken from an animal with a very strong built-up immunity so that the serum contains antibodies specific to the disease. The protection from this *hyperimmune serum* is only temporary. Foals acquire a similar temporary immunity from their mothers through colostrum.

Sometimes a situation demands both an immediate and a longer-lasting protection. This can be provided by a dose containing both hyperimmune serum and vaccine.

The invaders can be attacked externally by the use of both disinfectants and antiseptics (which are virtually synonymous since both reduce or destroy germs). Disinfectants tend to be stronger and more aggressive and thus are usually used to destroy germs on equipment or housing, as opposed to the body of the horse. The term 'antiseptics' tends to be reserved for products that may only stop germs multiplying, but they can safely be used on infected wounds as they will not inhibit healing.

An antibiotic is a chemical substance, such as penicillin, which kills bacteria or stops them multiplying. Sulpha drugs may also be used to inhibit the activity of bacteria and are used to treat some bacterial infections.

Parasites

Table 2.1(b) shows that disease may be caused not only by bacteria and viruses but also by fungi. These may be on the surface of the skin, e.g. ringworm, or elsewhere, e.g. on the lining of the lungs or other parts of the respiratory tract, or in the genital tract. The remaining parasites come from the animal kingdom and include acari (such as ticks and mites), together with insects (such as lice). Because these attack the outside of the horse they are called *ectoparasites*. Flies come within the insect group. They lay eggs on horses and the eggs hatch into larvae which continue their life cycle within the horse as *endoparasites*. The horse is also troubled with worms, including flatworms (such as liver fluke and tapeworms), and roundworms, the latter being the horse's greatest enemy.

Table 2.1 Some notable invaders: their causes and effects.

(a) Animal invaders

Disease	Cause	Symptoms
Ticks	Ticks	Irritation
Mange	Mites	Intense itching, scabs and loss of hair
Ear mange	Ear mites	Head shaking and restlessness
Autumn itching	Harvest mites	Restlessness, stamping, scabby legs
Lice	Lice (lice eggs = nits)	Irritation, possible hair loss, unthriftiness
Bots	Bot flies	Unthriftiness
Warbles	Warble flies	Swellings on the back in early summer
Fluke	Liver fluke	Unthriftiness, lack of growth, anaemia
Tapeworm	Tapeworm	Unthriftiness
Whiteworms	*Ascarids*	Lack of growth and lung damage in foals.
Seatworms	*Oxyris*	Irritation causing tail rubbing
Lungworms	*Dictyocaulus*	Coughing
Bloodworms	Redworms (*Strongylus*)	Colic, unthriftiness, anaemia, poor coat

(b) Microbe invaders

Disease	Cause	Symptoms
Ringworm	Ringworm fungus	Circular patches of raised hair, going bald
Broken wind	Fungus (an allergic reaction)	Cough, respiratory problems
Fungal abortion	Fungus	Abortion
Lockjaw	Tetanus bacterial toxin	Spasms, stretched-out stance
Strangles	Specific bacteria	Nasal discharge and abscesses between jaw bones and in neck glands
Poll evil and fistulous withers	Brucellosis bacteria	Infection of the relevant areas
Acne	Bacteria	Roughened skin with sores
Contagious equine metritis (C.E.M.)	Contagious equine metritis bacillus	Discharge from the vulva
Pneumonia	Virus infection (one of several possible causes)	Inflamed lungs, fever
Influenza (epidemic cough)	A specific influenza virus with three main strains identified	Cough, loss of appetite, nasal discharge and fever
Cold	Cold viruses	Cough and running nose
Swamp fever (infectious anaemia)	Specific virus	Fever, debilitation (*Note*: horses for export may have to be checked by Coggins test)
Warts	Papova virus	Small skin growths
Angleberries (sarcoids)	Papova virus	Nasty growths which tend to spread and bleed
Spots or pox (coital exanthema)	Herpes virus	Sores on vulva or penis

Symptoms of Disease

A symptom is an indication that something is wrong. It is a warning sign that should prompt investigation.

Behaviour

The first sign of disease is often a change in behaviour – anything which is different to that horse's normal pattern. The person in charge of the horse must ask himself the cause for the horse's change in behaviour. Is this attributable to external events, or to something in the creature's body or mind?

A mare about to foal, or a horse with colic, may display similar symptoms. Any abnormal activities should be noticed. The horse is normally alert and interested: any dullness or lack of zest should be regarded as a warning sign. It may be an indication of pain or of something else.

Appetite

A horse that does not hurry to the manger or finish a meal should always be regarded with suspicion. The horse may chew the food and let some slip back out, or it may have difficulty in swallowing. The cause may be in the food or in the manger, but it may be in the horse's mouth or in its digestive system. A stabled horse that drinks more or less water than usual should be similarly regarded, although it is to be expected that a horse will feel more thirsty on hot days or after sweating.

Action

When the horse is free in the field or turning round in the stable, or being ridden or driven, telltale signs may be evident. For example, the ears may suddenly flick back or the tail may be clamped down. The horse may grunt when mounted or be unwilling to go forward. Its stride may be uneven or it may be lame. It may rest a leg when standing in the stable but this is generally suspicious only if it is a front leg.

When the horse resists going forward, it is one of the most difficult decisions that the horseman has to make as to whether the animal is being stubborn or nappy, or whether it is in pain for some reason, or finds movement physically difficult or frightening. The horseman's great dilemma is to know when to punish and when to be understanding; when to encourage and when to be firm.

Coat

A harsh, 'staring', dull, tight coat is unnatural and is usually a sign that something is wrong. The coat should be soft and move freely over the muscles. Except when the horse is cold, the hairs should lie flat and the coat should gleam. Rough or raised patches, rubbed hair or any local differences in one area should be watched out for. The horse should also be checked for cuts, wounds, splinters and bruises.

Respiration

Changes in respiration may be noticed in the stable or the field. The respiration rate will rise during fever and infection. Respiration type is also significant; shallow and rapid breathing is characteristic of infections of the respiratory tract. Respiration rate will be affected by the environment, rising in hotter and more humid conditions.

There may be a cough when the horse is feeding or working, and it is important that the circumstances and type of cough should be noted to assist in the diagnosis. The horse may make a noise when galloping. It is important to note if the noise is made by air going into or coming out of the lungs, as explained in chapter 6.

Temperature

Whenever the horse is thought to be unwell, its temperature should be taken as this gives one of the most useful guides. An above-normal temperature accompanies all cases of acute disease to a greater or lesser extent. It will also indicate fever, a local infection, such as an abscess or one caused by the presence of a thorn, and pain, whether acute or general.

A fall in temperature is characteristic of loss of blood, starvation, collapse, coma, hypothermia and some chronic diseases. An abnormal temperature indicates that the vet should be called.

Pulse

The pulse rate is a useful aid to diagnosis and to help determine fitness. The pulse rate at rest rises in cases of fever and acute pain; it falls in debilitating diseases.

Dung and Urine

If the faeces are too hard, or too soft, strong smelling or slimy, all is not well within the digestive tract. Urine of unusual colour, cloudiness or smell may be a sign that problems are developing.

Eyes

A dull eye or one that is half closed is an indication that the horse may be feeling unwell. A special watch should be kept for damage to the eye. The inner side of the eyelid is a useful membrane to study, as it will change colour according to the condition of the blood. The gums may also be studied similarly. Healthy horses must be examined regularly so that any changes become apparent quickly.

Lumps, Bumps and Swellings

It is easy to find swellings on horses when grooming them; it is much harder to notice such things on animals in the field.

Swellings by the jaw bone may indicate glandular disorder. The inside of the mouth should be checked occasionally for ulcers and sores. When unaccustomed tack or clothing comes into use, particular care should be taken until the skin has hardened. A·sore or rub is more easily dealt with if detected in the early stages. Where there is swelling, it should be checked for heat, and bruising, strains, thorns and infection should be considered.

A watchful eye should be kept on the legs; the tendons that go down the back of each lower leg should stand out clearly. Slight filling, or puffiness round the fetlock joint, is a danger sign which must not be ignored. The cause might be a knock the day before, a strain, exercise on hard ground, or being shut in the stable too long. The first essential is to note the symptom; the second essential is to realise its significance.

Discharges

A runny nose is the common first symptom of a cough or cold but it may have other meanings, particularly if only one nostril is affected. Discharges may appear at any of the body's other orifices: eyes, ears, anus, vulva, sheath or teats. Each discharge will have its own particular meaning.

First Aid

If the problem is of a minor nature, it may well be dealt with using the equipment kept in the stable yard. The commonest troubles are cuts, small wounds and scrapes; and the first-aid kit is prepared with this in mind. The first-aid kit should be complete and readily accessible.

First-aid Kit

The best place to keep a first-aid kit is in the tack or feed room on top of the medicine chest, which should be kept locked. A second kit may be needed for travelling. As the human first-aid kit could to advantage stand beside it, the container for the horse's kit should be clearly marked, for example, 'First Aid – Horse'.

A clean bowl is often required, and so a useful container for the first-aid kit is a large plastic bowl with a well-fitting lid. The vet's name and telephone number should be clearly written on the inside of the lid. A clean cloth serves as a cover-all inside the bowl and is useful for covering a small table or straw bale when the contents are laid out. The kit might include the following, which should conveniently fill the bowl:

Bowl for antiseptic solution
Cloth upon which the kit may be laid
Antiseptic solution ready diluted in the travelling kit
Wound powder in a plastic 'puffer' bottle
Antiseptic cream, e.g. Acriflavine
Scissors with rounded ends and slightly curved blades
Forceps or tweezers, for removal of thorns
Thermometer (stub-ended, 30-second type)
Cotton wool substitute, a small roll
Gauze, a small packet
Elasticated bandage about 6 cm ($2\frac{1}{2}$ in) wide
Poultice (ready-to-use type in sealed pack)
Wound dressing in sealed pack
Money for a telephone call (in the travelling kit)

Expendable items must be replaced as used. The contents must be kept scrupulously clean. Many of these items may be purchased from the vet.

As well as the first-aid kit there should be a medicine cupboard. Its contents are largely a matter of personal preference but the following list may be helpful:

Polythene bags as plastic backing to poultices
Gamgee tissue (baby's nappy roll is also useful and cheaper)
Antiseptic aerosol spray, generally with purple tracer. (If this has an antibiotic base it will have a shelf life of only two years. Use with care as the horse may take fright or kick at the noise.)
Jar of udder cream to protect heels
Jar of cream for treating cracked heels and similar sores
Cough electuary
Worm powders or paste
Cooling lotion

Tin of kaolin poultice
Prepared poultice dressing pack
Fly repellent
Eye ointment & bottle of human eye wash
Astringent powder for leg paste
Working blister
Liniment
Dermicidal soap for skin infections
Crepe bandages – *not* open-weave cotton
Hoof-care preparation
Tranquillizer for use in an emergency (preferably in consultation
with the vet)

First-aid Procedure

Safety
The horse must be controlled quietly but firmly. It is best to have it
held by someone else and if it becomes fractious a bridle should be
fitted for greater control. A restraining influence may be obtained
by holding up a front leg. When examining a hind leg, it sometimes
helps to hold the horse's tail firmly downwards. A horse may also
be restrained by grasping a fold of skin on its neck. There is no
point in either the horse or a handler being hurt, and if necessary a
twitch should be used on the upper lip.

Calm
The attendants should talk reassuringly to the horse, pat and
stroke it and work without fuss or bother. This will maintain a
calm atmosphere. Thoroughness is more important than speed.

Blood Control
There are three types of bleeding or haemorrhage. First, there is
the blood around a cut from the tiny capillaries in the flesh; this is
not serious. The second type flows gently and is dark red; it comes
from the veins and is called venous bleeding. The third type is
bright red and comes from an artery; the blood runs freely and
may spurt out under pressure from the heartbeat. An injury
involving bleeding of the first type can wait for treatment until
return to the horse transport or stable. However, any injury
involving venous and arterial bleeding calls for immediate
treatment.

Venous bleeding can be controlled with a clean pad such as a folded handkerchief placed on the wound, and then securing the pad firmly over the wound using a bandage, tie, stock or belt.

Where arterial bleeding is dominant, the blood flow is more difficult to stop. As time is important, it may be necessary to apply a temporary tourniquet. To do this, a small pad is placed on the artery, just above the wound on the leg, and a tight restriction is pulled round it thus cutting off the blood supply. This can be achieved by using a loose tie and placing a stick in it: the stick is twisted, thus taking up slack, and is then secured. *As a tourniquet cuts off all the blood to the lower part of the limb, it should not be left on for more than a few minutes.* In this time it should be possible to get a good pressure pad on the wound; however, care must be taken to ensure that the securing bandage does not act as a tourniquet. If the wound is not on a leg, it may be necessary to hold the pressure pad in place until help can be obtained. The horse should be kept warm and quiet until it is taken home for treatment. If the need arises, help must be summoned or the horse led to the nearest house, where transport can be arranged. If any doubt exists, the vet should be called as the wound may need suturing, and antitetanus protection may be required.

Cleanliness

A dirty wound should first be washed under a cold hose, although care must be taken not to frighten the horse. Hair is then cut away from the region of the wound. A piece of cotton wool soaked in antiseptic solution is applied, taking care to wipe dirt out and not rub it in. Each swab should be used once only, and should not be put back into the disinfectant solution. When the wound is clean, it is dried with a dry piece of cotton wool. Wound powder is then 'puffed on' lightly. Antiseptic cream is useful for sores and grazes. Where a dressing is required to keep out dirt, the wound is covered with gauze, preferably medicated, and a cotton wool pad is bandaged gently in place.

Choke

First aid is concerned principally with preventing death from loss of blood or, in rare cases, from failure to obtain air, which is called asphyxia; for example, a horse breathing very deeply after severe exertion is given food, which gets stuck in its larynx thus causing rapid death. If a horse is choking, it will appear distressed and keep trying to swallow; saliva will run from its mouth or nose. In such circumstances asphyxia is unlikely. The first task is to identify where the obstruction is. If it is lower than the pharynx, at the back of the throat, it will not greatly interfere with air intake. It may be possible to feel the obstruction in the gullet on the left side of the neck just behind the windpipe, which is the tube down the front of the throat. Massage may be effective in moving the obstruction. The horse should not be drenched, or it will be drowned by the fluid going into the lungs. If the obstruction is at the back of the throat, by holding the tongue out to the side it may be possible to put a hand into the mouth and remove the obstruction, although there is a risk of being bitten. In other cases the vet should be called.

Fractures
Where there is a suspected fracture of a leg, the horse should be restrained. A horse with a broken leg can be a very distressing sight, and while a vet is being summoned onlookers should be kept back and something used to screen the horse from view.

Principles of Nursing

General
The first essential is to keep the horse comfortable and relaxed. If the horse is not getting exercise, the diet must be cut right back and should be such that it keeps the bowels working well. Cut grass can usefully be included in the feed, as can damp bran and other slightly laxative foods. Oats and barley should be avoided, but carrots and apples are beneficial. The horse should be kept warm with deep bedding, leg bandages and rugs as necessary, but with ample fresh air, without draughts.

If the horse breaks out into a sweat, it should be dried with an old towel and its ears should be 'stripped' by grasping them gently round the base and sliding the hand to the tip, working on each ear alternately.

The horse's drinking should be monitored and an automatic drinker should not be used. At feeding time, the sick horse should be dealt with last to avoid carrying infection. The horse's legs may fill if it is not being walked out. A thorough twice-daily massage to the legs will relieve this. Rubbing should be towards the heart, and stable bandages should then be applied immediately, bandaging upwards over padding. Provided its condition allows, the horse should be well groomed. A very sick horse may prefer a quick wipe over with a damp cloth, including nostrils, eyes, sheath and dock. When grooming, care should be taken not to let the horse get chilled. The top door may be shut and in winter an infra-red heat lamp may be used.

The instructions on all medications should be read with care and followed exactly. The vet's instructions should be carried out correctly too, so they should be carefully noted to avoid mistakes. As in hospital, a T.P.R. chart with clinical notes should be kept.

Hygiene is important at all times, but especially so for the sick horse. The grooming kit and feed containers must be kept scrupulously clean and should not be used for any other horse. The coat and feet should receive 'better than ever' attention. All things used in connection with a sick animal should be kept away from the equipment for the other horses. It is also wise to wear a clean smock or work-coat when working with the sick horse.

The sick box, like a foaling box, should be designed so that it can be cleaned and disinfected regularly and thoroughly. Sun and wind will disinfect, and so the sick box should be left open when not in use. Before disinfecting with chemicals, the building must be completely clean; it is then scrubbed with warm disinfectant solution. Creosote is a good disinfectant for unpainted wood. A useful disinfectant for concrete floors is a hot solution of washing soda. At least a day should be allowed before the disinfectant is washed off with clean water. Kit which is soaked in a disinfectant solution should be immersed for at least six hours. In 'disease-free' areas, great importance is always attached to clean footwear, so it is a sound idea to have a boot-wash situated where mud can be washed off rubber boots. There should also be provision for the attendants to wash their hands.

Isolation

All diseases associated with germs are infections and are infectious. The passage of the disease from one animal to the next can be made less likely by isolation procedures. There are two forms of isolation: within the yard for a contagious disease (one carried by contact), or outside the yard for airborne infection. The contact that spreads a contagious disease need not be direct. For example, a horse with a skin infection may be ridden under saddle, after which the saddle may be placed temporarily on a saddle horse. Later another saddle is rested on the saddle horse and may well pick up the germs and carry them on. Mildly contagious diseases can be kept under control by careful application of the principles outlined here.

Some diseases can be carried in the air. When a horse coughs, it releases germs which may float down wind. Birds and flies also carry germs. With more highly infectious diseases, the ideal is an isolation box sited about 400 m (around a quarter of a mile) down wind of the stable yard. This gives a better chance of keeping a disease out of the yard itself. Such a box can also be used for visiting horses and, for the first fortnight, for new arrivals in a yard.

Treatments

Cold Hosing

Where there is any bruising or tearing of the tissues, cold applications will shrink the blood vessels. To control swelling after injury, cold and pressure may be needed for the first day, but then heat is required to aid healing. Cold is easily applied by running cold water. While an assistant holds the horse using a bridle, the hose should be run very gently, first on the ground and then on the foot; it is then worked gradually up the leg. This should continue for ten minutes, and several sessions per day are needed, using pressure bandages between sessions. If the horse can be stood in a suitable stall, such as in a horsebox or trailer, the horse can be allowed to eat from a hay net and the hose can be bandaged to its leg. Alternatives to cold hosing include standing or walking in the river or the sea.

Cold Bandages, Massage and Astringents

An ice pack can be made by crushing some ice cubes in a cloth with a hammer or rolling pin, and then transferring this crushed ice to a polythene bag which may be bandaged on to the leg over a thin layer of gamgee to prevent skin scald. Methods of bandaging are shown in Fig. 2.2.

In addition to using cold bandages, some swellings – particularly those involving filling of the legs – respond well to massage. The legs should be rubbed upwards, towards the heart. To reduce

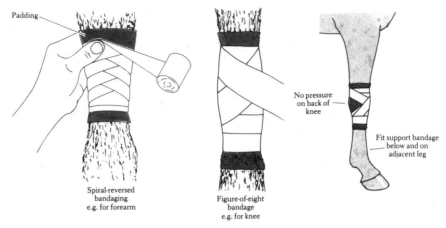

Fig. 2.2 Bandaging.

friction, it may help to use soapy water, baby lotion or oil. Alternating hot and cold applications also produces a massaging effect.

Some horses, particularly those that have to gallop or jump on hard ground, tend to get filled legs after work. This condition may be helped by using a cooling lotion after work, or by rubbing a diluted astringent liniment into the lower leg in the evening. As a general rule, bandages should not be applied over these liniments or the skin may be blistered. As a preventative measure against legs filling, there is a useful astringent which is purchased in powder form. This powder is mixed with water to form a paste, which is then applied with a spatula or spoon handle. The hair is wetted, then the paste is applied, first against and then with the lie of the hair. The paste sets so the horse can be worked at home with the paste on. The paste can be washed off when the horse is required to be tidy.

Poulticing

Where there is damaged tissue, the application of heat will stimulate the blood supply to the area and this will help repair the damage. A poultice also has a drawing effect, which will help any pus form into an abscess. It is thus often useful for a wound, whereas an ice pack is useful for a bruise, and an astringent for reducing swelling.

Impregnated padding can be purchased ready for use, complete

with instructions on the pack. Alternatively, kaolin or antiphlo-
gistine poultice may be used. The procedure is simple. The lid of
the tin is loosened and the tin is then placed in a pan of boiling
water for several minutes, until the paste is as hot as can be borne
on the back of the hand. Paste is then put on a piece of lint, and
covered with gauze, and is then applied. The poultice is then
covered with a polythene sheet so that it draws from the wound
and not from the air. The next cover is a piece of gamgee or
similar padding, to retain the heat, and finally it is bandaged in
place. The dressing is usually changed morning and evening.

If the wound or sore is in the sole of the foot, some people like
to use a bran poultice. Boiling water is mixed with antiseptic and
poured on to bran until a crumbly consistency is obtained: if the
bran is squeezed, there should be no excess moisture. The mixture
is allowed to cool until it can be tolerated by the palm of the hand
and then it is placed in polythene inside sacking. The foot is stood
in the bran and the wrapping secured around padding on the leg. It
is tidier to finish off with a stable bandage. The poultice should be
renewed night and morning. The antiseptic will mask the smell of
the bran so that the horse will not try to eat it.

A puncture wound is sometimes poulticed with a mixture of
Epsom salts (magnesium sulphate) and glycerine.

Irrespective of the type of poultice in use, there is one situation
which demands particular care. If an open wound lies over a joint,
there is a possibility that the joint capsule may be damaged and a
poultice could draw out the joint oil. A poultice should never be
used in such cases.

Where the horse is taking more weight on the sound leg than on
the injured one, this should be bandaged for support.

Fomentations

Fomentations are a useful way of applying heat to an area which is
not easily poulticed. They should be repeated several times a day
and continued for about 20 minutes on each occasion. A bucket
and a container of hot water should be taken to the horse. Hot and
cold water should then be mixed to a temperature which can just
be tolerated by the human elbow. A double handful of Epsom salts
may be added. A cloth, such as an old towel, is then soaked in the
water, wrung out and applied to the area for a couple of minutes.
This procedure is repeated, keeping the water as hot as can be
tolerated.

Tubbing

Open wounds in the foot may call for regular applications of heat; this can be achieved by tubbing. The preparations are similar to fomentations but a non-metal bucket or tub is used. Some antiseptic may be added to the water and as much Epsom salts as will dissolve, thus making a saturated solution. The horse's foot is placed firmly in the tub or bucket. The water may be hotter if the water level does not come above the top of the hoof. Generally, however, it is easier to have the water hand-hot. If the horse is reluctant to immerse its foot, hot water should be splashed gently over the leg until the horse is willing to lower it into the bucket. Tubbing should last about 20 minutes and should be repeated at least twice a day.

Steaming the Head

Where there is considerable discharge from the nostrils, the head may be steamed. To prepare for the treatment, a handful of hay is placed in a plastic bucket in the bottom of an old sack. This is then sprinkled with friar's balsam or oil of eucalyptus. Boiling water is poured over the hay so that it steams. The horse's head is put into the entrance of the sack and kept there for several minutes. More hot water is then poured over the hay and the procedure repeated. From time to time the horse will need a break. For heavy catarrh it may be necessary to steam the head twice daily.

In cases of pneumonia the sack should not be used as it limits fresh air. After the steaming process, the hay will be contaminated with nasal discharge and should be burned. The bucket and bag must be scalded to sterilise them.

Horses being steamed are best fed at ground level to encourage discharge. A little ointment containing menthol, oil of eucalyptus or something similar may be placed in the outer nostril. If the discharge tends to create a sore, the skin should be protected with petroleum jelly or nappy-rash cream.

Electuary

This old-fashioned paste still provides a useful aid to soothing a cough. It is usual to draw out the horse's tongue and, using the flat handle of a spatula, place the paste on the tongue. The treatment is repeated at least twice a day.

Drenching

Giving a horse liquid medicine can prove to be a difficult business. Some horses accept a drench quite easily, but others are not so co-operative, so it is wise to take precautions. The horse's head should be raised so that the liquid can be poured down its throat. This is not difficult with a small pony but most people find their arms are too short to treat a horse this way. A rope should be attached to the middle front of the noseband of the head collar and the rope is passed over a beam. Extra height is gained by standing on a straw bale or two. This is essential if a beam is not available. An assistant can then control the horse and raise and lower its head as required. The drench is best placed in a plastic bottle. If a glass bottle is used, then the neck of the bottle must be bandaged in case the glass is broken by the horse's teeth. The animal's head is raised gently and the neck of the bottle is placed in the corner of the mouth. Gentle pouring may then commence. When one mouthful has been swallowed, another may be given. If the horse coughs, lower the head at once. If the medicine goes down 'the wrong way', it will end up in the lungs and may give the horse pneumonia. *Drenching should be done only by someone competent because of the risks involved.*

Other Methods of Giving Medicines

Some powders and granules may be added to the food, but the horse has a sensitive nose and palate and may easily detect these additions and be put off. The meal should therefore be made particularly tempting by the addition of apples or carrots. If the horse fails to eat all of the prepared meal, the value of the medicine might be lost. Some medicines can be poured straight into the water bucket.

A useful way to ensure that the horse receives the appropriate medicine dosage is to make up a paste with icing sugar to which the medicine is added. Using a large syringe, without a needle, the mixture is squirted into the horse's mouth towards the back of the cheek. Some drugs are now supplied as a paste in a syringe for use in this way.

Some treatments require administration to the skin, and for some of these, rubber gloves are required. In every case the great essential is to read the instructions or to follow the vet's directions.

Injections

It is now common for stockmen to inject certain classes of stock. The horse is not easy to inject as it has a tough hide and also sometimes produces a reaction to the injection. Furthermore, if the injection is not done smoothly and at the first attempt, the horse is apt to become fractious and be difficult on the next occasion when injection is attempted.

Injections may be given intravenously (into the vein), and this is certainly a job for the vet. They may also be given intramuscularly. If the vet approves of an experienced person giving intramuscular injections, then he can demonstrate the proper technique.

Enemas and Back-raking

It is sometimes helpful to the passage of faeces through the rectum to flush or lubricate this part of the bowel with a fluid. There are several other reasons for giving an enema, and different fluids are used for different causes. The normal procedure is to lubricate the rounded end of a special tube and pass it through the anus into the rectum. Then, soapy warm water or liquid paraffin is passed down the tube. When dealing with a foal, gravity is often used, but in the case of a horse, a pump may help to pass the fluid along the tube. The giving of an enema is usually a task for the vet.

Back-raking is the removal of faeces, or of meconium in the case of newborn foals. In the horse or pony, a well-lubricated hand is used, but for a foal a small, smooth, well-lubricated finger is all that there is room for without causing inflammation. Generally, back-raking is best left to the vet.

Treatments using Tubes

A stomach tube is used to place either a large quantity of fluid, or a small quantity without wastage, into the stomach. The tube is about 3 m (10 ft) long and about 12 mm (around half an inch) in diameter. It is passed up one nostril and goes down the gullet (oesophagus) towards the stomach. If the tube should accidentally go the wrong way at the pharynx, then it would pass into the trachea, and unless immediately remedied, the fluid would pass into the lungs. *The stomach tube is only used by a vet.* Occasionally, there will be minimal bleeding at the nostril, but this is of little consequence.

Where it is necessary to empty the bladder, and the horse seems unable to oblige, the vet will pass a thin tube up the urethra into the bladder and thus allow the urine to escape. Such a tube is called a catheter.

Tooth Care

Until the horse is four years old the teeth should be examined twice a year to make sure that the milk (deciduous) teeth are not getting in the way of the permanent teeth. After the age of four the teeth should be examined once or twice a year for sharp edges. To examine the molar teeth, it is easier to use a gag as the horse has sufficient power to damage a finger should the grip on the tongue be lost. However, care must be taken not to get the horse alarmed or excited. Sharp edges on both the top and bottom molars need to be rasped off with a long-handled tooth rasp. Most horses do not mind this operation, which can be done by an experienced operator instead of the vet, provided that the correct equipment is available, that reasonable care is taken and that the operator makes sure that the desired result is achieved. Rasping is a two-person job. The horse, in a head collar, is backed into a corner to face the light, and the assistant, standing on the opposite side to the person doing the rasping, steadies the head and may be required to hold the tongue. The rasp must be dipped into water at regular intervals to keep the cutting edges clean.

Aids to Diagnosis

A nerve block is effected by a vet injecting anaesthetic over a nerve to achieve a loss of sensation in the area covered by that nerve. This process can be used to achieve local anaesthesia for treating a wound or it can be used to help diagnosis. If a lame horse cannot feel its near fore foot and yet still limps, the seat of lameness is not in that foot.

Radiation is sometimes used to check inflammation and treat skin diseases. It is most often used to provide X-ray plates (radiographs), which are particularly useful in helping diagnosis of leg and foot problems. X-ray units with sufficient power to produce a satisfactory plate of the horse's body are only found at major equine research and teaching centres.

When a vet is diagnosing the cause of a problem, he may refer to the horse using terminology that indicates the aspect with which he is concerned. The terminologies used in the equitation and veterinary contexts are explained in Fig. 2.3.

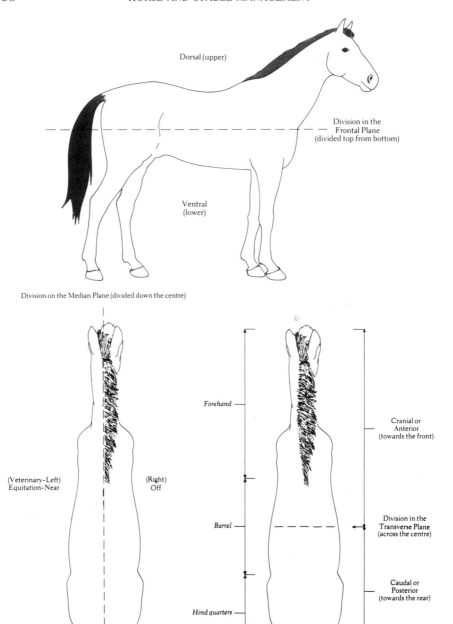

Fig. 2.3 Terminology.

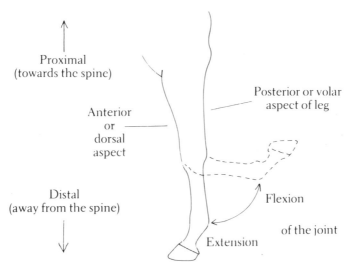

Proximal
(towards the spine)

Anterior
or
dorsal
aspect

Posterior or volar
aspect of leg

Distal
(away from the spine)

Flexion

of the joint

Extension

Fig. 2.3 contd.

Minor Operations

Castration is carried out either by giving a sedative and a local anaesthetic, which leaves the horse standing, or under a general anaesthetic, which collapses the horse on to suitable clean soft ground. When a horse goes down from a general anaesthetic, it is helpful to have a strong assistant to hold the leadrope and steady the horse. When the horse comes round, it may be easily upset and it is best left alone to struggle in peace, but the surroundings must be soft so that it does not hurt itself. After castration, there will be some swelling. However, the wound is free draining so, as long as the patient continues to eat well, there is usually no cause for concern.

Another common minor operation is Caslick's or 'stitching a mare'. When a breeding mare has poor conformation of the vulva, this may allow air and germs to enter the genital tract. This operation involves sealing the top half of the vulva under local anaesthetic by making two cuts and sewing them together. In spite of the adjacent position of the anus as a source of infection, this operation is usually trouble free.

Another simple routine operation is the removal of a tooth. Deciduous or milk teeth which have failed to clear out of the way for the permanent teeth are easily removed. To remove wolf teeth (see chapter 8), a local anaesthetic may sometimes be needed. For a large tooth it might be necessary to give a general anaesthetic. Horses vary considerably in the amount of interference they will tolerate without sedatives. It is helpful if those who are accustomed to the horse can advise the vet concerning its temperament and any relevant idiosyncrasies.

Treatment for Lameness

Lameness due to soft tissue may first be treated by the application of cold to reduce swelling, and later this may be interspersed with heat to draw blood to the area to aid healing. Heat may come from poultices, fomentations or ultrasonic therapy. Once the initial swelling has gone, it may be thought advisable to produce further inflammation by the use of counter-irritants. The simplest of these are known as 'blisters'. The first task is to clip the coat over the area. The blister is then applied according to the directions on the container. When blistering a tendon, the rear lower part of the leg should be covered with petroleum jelly or lard so that the blister does not run down the leg and inflame the heels. When blistering a joint, its inside angle should be avoided. The horse will need a cradle put on its neck so that it cannot nibble or lick the blister. Similarly, if the hocks are blistered, the tail must be kept bandaged up double so that it cannot swish over the blister and carry it to the flanks. The horse must be put on a very light laxative diet before the operation and for a few days afterwards.

Less severe treatment is given by the use of 'working blisters'; these do not always need the hair to be clipped. Generally, they are rubbed in with a soft toothbrush each day until the skin becomes scaly. A working blister is normally to be found in most medicine chests, whereas the use of a full blister is a matter for the vet, who will advise on current thinking of its value.

When the tendon is injured, the vet may decide that 'firing' (using a hot iron) is the most suitable treatment, and that both legs should be treated. The electrically heated iron may be used to make a row of deep but small burns. This is called 'pin firing'. Alternatively, the burns may be shallower 'lines', which leave less noticeable scars. Firing can be done using an acid instead of a hot iron. As an alternative to firing, some vets wound tendons in the 'split tendon operation' to gain extra support tissue. However, both firing and the split tendon operation have come under attack, and the veterinary profession is divided as to their usefulness. Another possible line of treatment is 'carbon-fibre implant', which is still being researched.

Whatever method is used to strengthen the legs, it is important to remember that the horse kept in the stable without exercise needs a light laxative diet, and that the sound leg, on the other side to the damaged one, will be taking extra weight and so will need support bandages. Eventually, and it may be after several months, the horse is turned out to grass. It will tend to gallop about and may reinjure the damaged leg. The front shoes should be left on for a front-leg injury but the back shoes can be taken off and the feet cut back so that the horse is very footsore behind and thus not inclined even to trot. Drugs can provide an alternative restraint, which can be used for the first few days. It is essential that the horse be left out at grass for some months, as complete rest is the best cure for such injuries.

Diagnosis of Lameness

The first task in the diagnosis of lameness is to decide which leg is the lame one. Lameness may be due to back problems, but generally the cause is in the lower leg.

The horse should be observed in the stable. If it points a fore leg, then that is probably the limb giving pain. The horse should be turned to left and right; if it drops on one side, then it is trying to favour the other side. The horse is then walked up and back and then is trotted up and back on a level surface. Its head jerks up when the lame front leg carries weight. The horse should be equally balanced so that at the walk and trot the weight should move evenly, with each limb taking an equal proportion to its neighbour. If the horse tries to save one limb so that it carries less weight, it may do so by using the head and neck as a counterbalance. To favour a leg means to reduce its work load by transferring weight off it quickly.

If both front legs are lame, the stride will be short and inhibited as the horse will try to take a shorter stride on an unsound limb. Less commonly, a back leg is the seat of trouble and is more difficult to diagnose. A left (near) hind causes the horse to carry its left hip up high but the head will drop as the lame hind limb takes the weight. Next, the horse should be worked on the soft to see if it runs up more sound. As shown in Table 2.2, a number of tests will help locate the lameness.

Once a decision has been made as to which leg is the lame one, it must be inspected more carefully. Good light is needed. The

Table 2.2 Tests to locate lameness.

Location	Tests and symptoms
Foot	Points at rest, warmer than the others, more lame on hard ground, heat at coronet
Tendon	More lame on soft ground, swelling and tenderness over tendon
Splint	Lameness comes on at exercise, lamer on hard ground
Knee	Swelling, heat, pain on flexion, followed by increased lameness
Shoulder	Move the leg and watch for pain symptoms
Hock	Limps on turning, lamer on hard ground. Hold up hind leg to belly for 30 seconds and then release, trot horse away and look for lameness. This is called the 'spavin test'
Stifle	Leg rested forwards, reacts to manipulation
Hip	Reacts to movement of the leg

knee should be pressed gently all round (palpated). Then the leg is picked up. The area down the length of each splint bone must be squeezed firmly but evenly. Following this, the same treatment is given to each major tendon and ligament, working carefully from knee to fetlock joint. If the horse lays back its ears, pulls its leg away, flinches or reacts in any other way, then it may be that the trouble is in that area. Until the trouble is located, the search must proceed, using sight as well as feel. The fetlock must be examined and then the heels. Heat or a sore place may be the only clue. Infection may enter through a tiny scrape. Every test must be compared with the same test on the neighbouring leg.

The foot must be examined, but it must first be picked clean and scrubbed with water. Observation may reveal damage on the sole. Each nail should be tapped with a hammer and pressure applied to the seat of corn. Shoeing history may be relevant. If the cause is not found, then it could be that the wrong leg was selected or that the problem lies above the knee.

Diagnosis of lameness is difficult and the experience of the vet may be needed; however, good observations may aid his task. 'No leg, no horse.' This is not a matter to be treated lightly.

Part II
The Systems of the Horse

3 Systems of support and movement

It is a characteristic of the higher orders of the Animal Kingdom to have a framework that gives structure and form to the body. The parts of a horse's body are fixed to a frame, the skeleton, which is built of bone and cartilage for strength. The main supporting member of the frame is the backbone which, together with the skull, affords protection to the central nervous system. The ribs also give some protection to the vital organs. There are also joints, bonded by ligaments, and muscles attached at one end by tendons to the bones in order to move them. The most vulnerable part of the system is the lower leg and foot, which occupies all of the attention of the farrier and much of the vet's equine practice. The task of these systems is to provide support, protection and movement – particularly locomotion.

Bones and Cartilage

The horse's skeleton consists of over 200 bones. These are so arranged that they can be used as rigid supports or become freely movable when the joints are brought into play, thus acting as levers to provide movement. Bones also store minerals and contain the marrow which is responsible for the formation of blood cells.

Bones are living structures with blood vessels. They are made of protein, (giving them strength) and mineral matter (which makes them hard and strong). There is less mineral matter in the bones of young animals, whose bones are softer in consequence. The mineral matter consists largely of calcium and phosphorous and this is why it is important to look closely at the ratio of these two minerals in the horse's diet.

Typically, a long bone, e.g., the cannon bone, is made of compact dense bone on the outside, forming the cortex. Within the

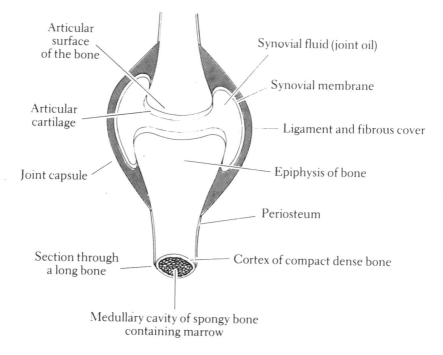

Articular surface of the bone

Synovial fluid (joint oil)

Synovial membrane

Articular cartilage

Ligament and fibrous cover

Joint capsule

Epiphysis of bone

Periosteum

Section through a long bone

Cortex of compact dense bone

Medullary cavity of spongy bone containing marrow

Fig. 3.1 Bone and joint.

cortex is the medullary cavity composed of spongy (cancellous) bone forming a porous network containing marrow. The bone is surrounded by a sheath of tissue called the periosteum, the cells of which are responsible for the formation of new bone material when young bones are increasing in girth. Bones grow in length at the epiphyses, which are areas just behind the end surfaces of the bone. The end surfaces where one bone meets another are called articulatory surfaces, and they are covered in cartilage.

Cartilage has a firm but slightly flexible consistency, providing a very smooth surface. When it is found in meat it is called gristle. Non-articular cartilage can be converted to bone by laying down minerals within the tissue, a process called ossification.

Joints

Bones meet at joints (see Fig. 3.1), which are of three types:

(1) *Immovable joints*, e.g. the junctions of the bones in the skull.

(2) *Slightly movable joints*, e.g. the junction between bones forming the spinal column or backbone.

(3) *Freely movable joints*, which take various forms: there are *hinge-type* joints, such as the fetlock; *plane-type* joints, such as the knee, where bones with flat surfaces glide over each other; *pivot joints*, which allow turning, e.g. the joint between the top bone in the neck (the atlas vertebra) and the second bone in the neck (the axis vertebra); there are also *ball-and-socket* joints, such as the hip.

Each freely movable joint is enclosed in a capsule, the lining of which is the synovial membrane. This secretes synovial fluid or 'joint oil', which acts as a lubricant. The outer cover of the capsule is fibrous and acts like a ligament, holding the joint together. There are similar capsules on prominent bone ends (such as the elbow), and these are called bursae (singular, bursa).

Ligaments are strong connective bands of tissue holding joints together. Some are within the joint capsule itself, but most are outside it, connecting the two bones on the sides, front and back of the joint.

Skeleton

The horse's skeleton is made up of bones, cartilage, joints and ligaments. It consists of the axial portion, made up of skull, backbones and ribs, and the appendicular portion, which is made up of the legs.

The Axial Skeleton

The axial skeleton is shown in Fig. 3.2. *The skull* consists of many small bones fused together to form protection for the brain, optic nerves, inner ears and nasal passages. One of the largest bones in the horse is the lower jaw-bone or mandible, which is hinged between the eye and the base of the ear. The skull also contains the teeth. The back of the skull is formed by the occipital bone, which has a junction with the top bones of the neck.

The neck contains the atlas, the axis and five other cervical vertebrae. The bones of the backbone or vertebral column are called vertebrae (singular, vertebra).

The chest part of the vertebral column contains 18 thoracic

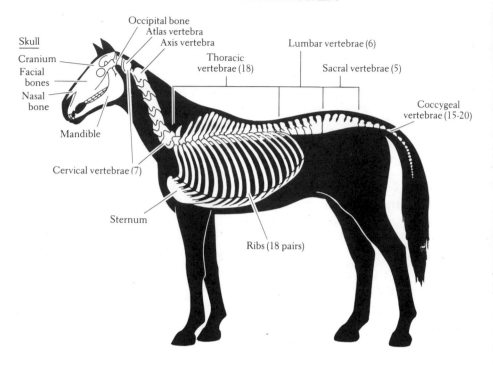

Fig. 3.2 Axial skeleton.

vertebrae to which are attached the 18 pairs of ribs. There are eight pairs of 'true' ribs attached directly to the breast bone or sternum. In addition there are ten pairs of 'false' ribs, attached to the breast bone by long cartilage extensions. The horse has no collar bone.

The next part of the vertebral column consists of six lumbar vertebrae forming the loins. Behind the loins are the five sacral vertebrae fused together as a firm base for the pelvis and forming the croup. There are also some 15 to 20 bones called the coccygeal vertebrae, which go down into the dock of the tail.

The neck and tail vertebrae are freely jointed to provide a wide range of movement, but little movement is possible through the thoracic and lumbar vertebrae. When we speak of a horse 'bending its back', most of the movement is in fact occurring in the neck. The apparent lateral bending is effectively a combination of neck, shoulder and limb alignment, the backbone remaining almost straight. The ribs limit movement through the barrel of the horse, and so any movement is confined to the region of the loins.

The vertebrae have vertical and transverse processes which aid muscle attachment. There is a canal or channel through the centre of the vertebrae housing the spinal cord. This is the continuation from the brain down the backbone. Occasionally, vertebrae seem to get slightly out of alignment and in some cases manipulation appears to help. Sometimes vertebrae will fuse together at their extremities. This can cause pain and reduced performance until the fusion is complete. Given time, however, the horse appears none the worse when the fusion is complete.

The Appendicular Skeleton

Figure 3.3 shows the bones that comprise the appendicular skeleton. The front legs are not joined by any bony attachment to the horse's axial skeleton. Thus, the weight of the horse is taken at the front by muscles, tendons and ligaments from the two front legs, so forming a sling in which the body is carried. This arrangement is part of the shock-absorbing mechanism built into

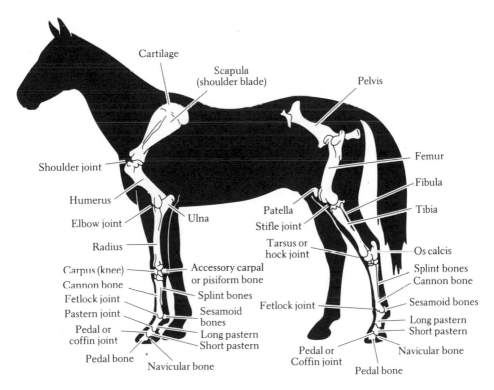

Fig. 3.3 Appendicular skeleton: the limbs.

the front legs. The other parts of this mechanism are the angles in the shoulders and fetlocks, the many bones of the knee, and the design of the foot itself.

The fore leg starts with the scapula or shoulder blade, the top part of which consists of cartilage. The scapula forms the shoulder joint with the humerus or upper arm bone. The front of the humerus is known as the point of the shoulder. The rear or distal (lower) end of the humerus forms the elbow joint together with the radius and ulna, which are fused together to form the forearm. The radius is the main weight carrier of the two bones, and the ulna reaches up to form the point of the elbow (olecranon process). Below the forearm comes the knee or carpus, which is the equivalent of the human wrist. It is made up of two rows of small bones with a small extra bone at the back, called the accessory carpal or pisiform bone. This little bone makes the tendon pull at an angle to bend the knee.

Below the knee are the three metacarpal bones; the central large one is the cannon bone, and the two small bones on either side of it are the splint bones. The cannon and splint bones are the equivalent to the three bones running across the back of the human hand (see Fig. 3.4). The bones equivalent to those of the

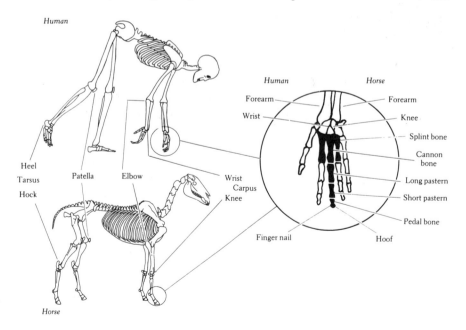

Fig. 3.4 Comparison of equine and human skeletons.

little finger and thumb have disappeared through evolution.

The cannon bone meets the digit at the fetlock joint. This digit is equivalent to the human middle finger. It consists of three principal bones (the phalanges): the long pastern (first phalanx), the short pastern (second phalanx) and the coffin or pedal bone (third phalanx). The joint of the long and short pastern bones is the pastern joint. That between the short pastern bones and the pedal bone is the coffin joint.

At the back of the fetlock joint there are two small bones designed to gain mechanical advantage in bending the joint: these are the sesamoid bones (proximal sesamoids). At the back of the coffin joint is another sesamoid-type bone called the navicular bone (distal sesamoid).

The hind leg starts with the pelvic girdle (see Fig. 3.5). This is made up from the fused sacral vertebrae and the two hip or pelvic bones (os coxae), which meet underneath at the symphysis. Each hip bone is formed from three bones fused together: the ilium, the ischium and the pubis. The ilium joins the sacrum on either side at the sacro-iliac joint. The front end of the ilium has the point of croup (tuber sacralae) near the sacrum and the point of hip (tuber coxae) on the outer side. The ilium unites with the ischium at the

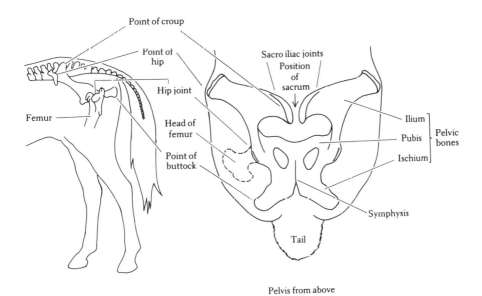

Fig. 3.5 The pelvis.

hip joint. The ischium has a rearward projection called the seatbone or point of buttock (tuber ischii). The pubis also meets at the hip joint, and the pubis, together with the lower part of the ischium, form the floor of the pelvis. Thus, the pelvis forms a complete hoop of bones protecting the vital parts of the horse's body. A long flat pelvis is said to be best for high speeds.

The thigh bone (femur) comes from the hip joint and runs down to the stifle joint. At the stifle there is another sesamoid bone called the patella, which corresponds to the human kneecap. The femur meets the tibia, which is within the second thigh or gaskin. In human beings, beside the tibia is the fibula; the fibula in the horse is only a vestigial remnant, sometimes no more than 10 cm (4 in) long. The tibia goes to the hock (tarsus), which, like the knee, consists of several small bones. The hock equates to the human ankle. At the back of the hock is a bone called the calcaneus or point of hock (fibular tarsal), which acts as a lever to extend the leg. It forms the heel in man. Below the hock, the bones are similar to those below the knee.

Muscles

There are three main types of muscle in horses:

(1) *cardiac* muscle, which is found in the heart.
(2) *smooth or involuntary* muscle, which is generally found in automatic systems such as the walls of the digestive tract.
(3) *skeletal* muscle, the flesh of the horse, which is like the red meat that we eat.

Each fibre of a muscle is controlled by a branch from a nerve. These muscle fibres are arranged in bundles surrounded by connective tissue. When stimulated, muscles reduce in length, thus exerting a pull. The more muscle fibres involved, the stronger is the pull. Thus, for extra strength the horse must build extra muscle. Muscles are arranged in sheets and bands, and in herring-bone and spindle-like groups according to their function.

Each muscle has an origin where it is attached to a stable part of the skeleton. At the other end it has an insertion into the part of the skeleton which it moves. Where the bone to be moved is distant from the muscle, there is a dense fibrous connection between them called a tendon. Tendons are usually cord or band shaped, but some are flat like sheets. They have little elasticity and are poorly supplied with blood. They also take considerable strain and, because of these factors, take a long time to heal when injured.

Just as a joint has a covering that supplies lubrication to help the moving parts, each tendon has a sheath where it passes over a joint, which protects and lubricates it. Similarly, just as a prominent bone end has a protective capsule (bursa), there are tendon bursae to help tendons pass over bones at a joint.

If a joint is to be bent, flexor muscles pull and extensor muscles allow. Other muscles (fixator or synergic muscles) steady the rest of the body. When the joint is to be straightened, the extensors are directly responsible muscles (agonists or activators) and the flexors (antagonists) allow or give while the limb is steadied by the other muscles.

When a muscle acts through a tendon on a bone to operate a joint, it gains mechanical advantage in different ways, as shown in Fig. 3.6.

The muscles of the horse's body are too complex to name and separate in a simple study. This is because there are many overlapping layers and the shape of the outer, superficial muscles is partly dependent on the deeper muscles. Some of the main muscles

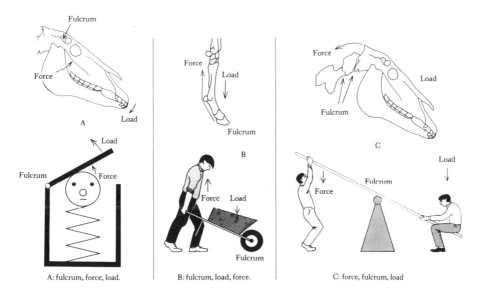

Fig. 3.6 Bones as levers.

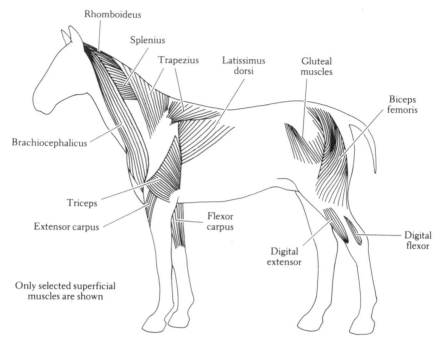

Fig. 3.7 Muscles.

of interest to horse-owners are as follows and as shown in Fig. 3.7.

Trapezius, the muscle on either side of the withers. It lies over the rhomboideus and splenius muscles. In some horses these muscles are poorly developed, which gives them prominent withers and 'ewe necks'.

Brachiocephalicus, the muscle which pulls the shoulder forward and is attached at its front end to the back of the head. This is why it is easier for a horse to jump well using its shoulders if it is given sufficient length of rein to extend the neck and head in flight. Similarly, at the collected paces where elevation of the steps is required, this muscle helps to carry the neck high and not stretched out, thus raising the shoulder.

Latissimus dorsi, the muscles from the shoulder blade to the back. Those running along the back are the *longissimus dorsi*. These are the muscles on which the rider sits.

The elbow has extensor muscles called *triceps* and flexor muscles called *biceps*. The lower joints also have paired muscles, but in order to keep the lower leg light they are kept in the upper leg. This is advantageous for high speed. The extensor muscles acting on the hip are known as the hamstring muscles. These pass up the back of the hindquarters and attach to the croup. On fit horses these muscles stand out clearly with divisions between them.

By removing a minute portion of tissue (in a biopsy), laboratory analysis has shown fast-twitch and slow-twitch muscle fibres. The mix of these in a racehorse has bearing on its sprinting or staying ability.

The Lower Leg and Foot

A man wearing heavy boots is slower than a man in lightweight running shoes. Evolution favours the fast-running horse to escape from its enemies, and similarly selects the horse with lightweight lower legs and feet. The remote ancestor of the horse had several toes; the modern horse has only one, which takes all the strain and is sometimes the weakest link. This is particularly so in the front leg, which takes all the strain on landing from a jump and which normally carries about 60% of the horse's weight.

The main strain is taken by the suspensory ligament (see Fig. 3.8) coming from the back of the knee and running against the cannon bone down the back of the leg to the fetlock. Part of the suspensory ligament is then attached to the sesamoid bones and part divides into two and comes round the pastern from each side.

The two tendons running down the back of the lower leg are together called the superficial flexor tendon and under it is the deep flexor tendon, which has a check ligament. This takes some of the strain from the muscles situated above the knee in the forearm, or above the hock in the second thigh. The deep flexor tendon runs over the sesamoid bones down to the pastern, into the hoof, and round the navicular bone. It is attached to the coffin or pedal bone. The superficial flexor tendon divides into two branches at the fetlock: these attach on both sides to both pastern bones.

The extensor tendons run down the front of the leg. As they take no weight, they are slim and generally trouble-free.

There is a band of tissue at the coronet called the coronary band. This creates horn and so produces the hoof. The outer layer of the hoof, called the periople, acts like a varnish to keep the moisture in. There is next a horny layer and inside that the insensitive laminae. These mesh with the sensitive laminae that surround the pedal bone, and in this way the pedal bone is firmly held in the foot.

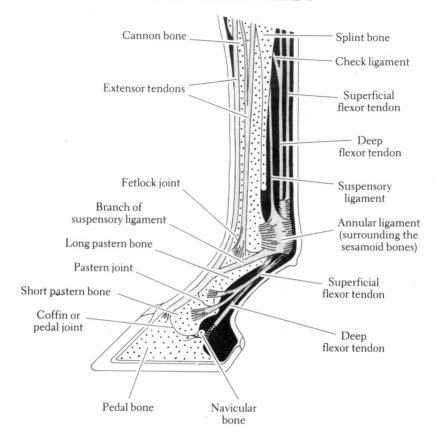

Cannon bone

Splint bone

Check ligament

Extensor tendons

Superficial
flexor tendon

Deep
flexor tendon

Fetlock joint

Branch of
suspensory ligament

Long pastern bone

Pastern joint

Short pastern bone

Coffin or
pedal joint

Suspensory
ligament

Annular ligament
(surrounding the
sesamoid bones)

Superficial
flexor tendon

Deep
flexor tendon

Pedal bone

Navicular
bone

Fig. 3.8 The lower leg.

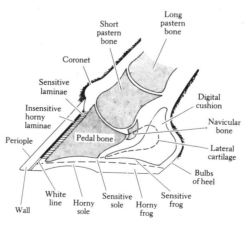

Short
pastern
bone

Long
pastern
bone

Coronet

Sensitive
laminae

Insensitive
horny
laminae

Periople

Pedal bone

Digital
cushion

Navicular
bone

Lateral
cartilage

Bulbs
of heel

Wall

White
line

Horny
sole

Sensitive
sole

Horny
frog

Sensitive
frog

Fig. 3.9 Foot structure.

The lower surface of the foot is composed of the sole and the raised frog and bars, which act like the treads on the wheels of a tractor, to help grip the ground. The shape of the underside of the foot is slightly concave, so aiding grip.

Under the back of the pedal bone is the pedal (plantar) cushion, which takes and spreads some of the weight from the short pastern bone; it does this by pushing wide the lateral cartilages, two wings of cartilage attached to the pedal bone, so spreading the heel and pressing the frog against the ground. Much of the concussion of normal working is thus absorbed within the foot. The foot is well supplied with blood and the action of the pedal cushion being compressed at every step is like a small pump helping the circulation.

Figure 3.9 shows the structure of the foot.

Farriery

A normal hind foot is longer than the fore foot. When the horse is working under stress, it is important to keep the toe of the fore foot short, the frog in contact with the ground and the heels wide. This helps the foot to work most efficiently.

The rate of the growth of horn varies according to food, environment and exercise, and is greater at the front of the foot than at the heel. Wild ponies normally walk far enough and live in sufficiently rough conditions to keep their feet correctly worn. An unshod horse in a field may need its feet rasping every six weeks, as will a shod horse, although in the latter case wear of the shoe may dictate more frequent treatment.

If the farrier is attempting to alter action by trimming the foot, this must be done only a little at a time, as it will alter the angle of wear on the joints. The foot is usually trimmed so that the angle of the pastern to the ground is the same as the angle of the foot (see Fig. 3.10). When rasping, it is harder work to take down the front of the foot and there is a tendency to take off too much at the heels or to reduce the bars, which may lead to contracted heels.

The preparation of the hoof for the reception of a shoe consists of removing loose particles of sole and tidying ragged pieces of frog. The bearing surface of the wall must be made level with the outer edge of the sole.

Although a shoe may not be worn out, it must be refitted every four to six weeks, otherwise it will be carried forwards by hoof growth. The weight of the horse borne by the shoe should not be

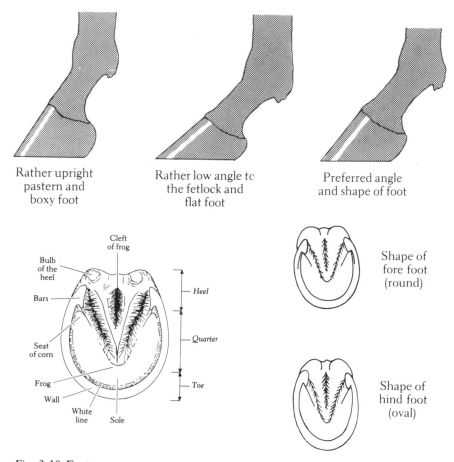

Rather upright pastern and boxy foot

Rather low angle to the fetlock and flat foot

Preferred angle and shape of foot

Cleft of frog

Bulb of the heel

Bars

Seat of corn

Frog

Wall

White line

Sole

Heel

Quarter

Toe

Shape of fore foot (round)

Shape of hind foot (oval)

Fig. 3.10 Feet.

far in front of the line down through the centre of the cannon bone, otherwise damage to hooves, joints and tendons will occur.

Good farriers can improve horses' feet and action. The specialist work of this highly trained, skilled craftsman, such as corrective shoeing, can only be adequately discussed at length and therefore is outside the scope of this book.

Disorders of the Skeletal, Articular and Muscular Systems

The main disorders (Fig. 3.11) of the systems of support and movement are as follows.

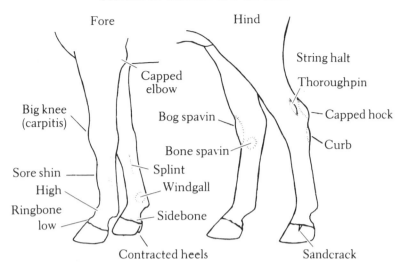

Fig. 3.11 Leg problems.

Abscess in the Foot

Symptoms: Lameness caused by pus in the foot. Heat and tenderness. (Gravel is a similar condition.)
Causes: A penetrating wound or a bruise.
Treatment: A hole must be cut to allow the pus to drain. Tub the foot, and poultice twice daily until the condition is cleared. The hole must then be plugged securely.

Arthritis

Symptoms: Heat, swelling and pain at a joint.
Causes: Inflammation of a joint from any cause, e.g. an infection, a blow or a sprain.
Treatment: Rest, together with treatment of any infection. (*Note*: the use of drugs such as phenylbutazone ('Bute') may mask the pain. Such drugs are anti-inflammatory.)

Bruises in the Foot

Symptoms: Lameness and pointing or resting the foot. Pus may break out at the heel from a septic corn. The horse flinches from pressure or when the area is tapped. Corns – at the seat of the corn. Bruised frog, sole or heel.
Causes: A sharp stone, etc., or, for a corn, a badly fitting shoe.
Treatment: For a corn remove the shoe. Thin the sole or seat of the

corn, cutting down towards the reddened area. Tub the foot, and poultice, repeating twice daily for two or three days. Re-shoe with care, possibly using a seated-out or three-quarter shoe if there is a corn.

Bursitis

Symptoms: Soft swelling at the bursa, sometimes with initial heat and pain. This condition is called synovitis if a tendon sheath is involved.

Causes: A knock, e.g. through travelling in too short a space, may cause a capped hock. Using too little bedding may cause a capped elbow, which may also be caused by a knock from the heel of a front shoe. A knock from a jump pole may cause a big knee (carpitis). Bursitis at the poll may result from a knock on a low roof. Going over backwards or a poorly fitting saddle may cause bursitis at the withers. Bog spavin, which is a soft swelling at the front of the hock, may be caused by a strain or injury. Concussion, e.g. too much ' 'ammer, 'ammer, 'ammer' on the hard road, may cause windgalls. Thoroughpins may be caused by a strain, e.g. jumping an unfit horse in deep going.

Treatment: If a capped elbow is caused by the heel of a shoe, then a protective ring, called a 'sausage boot' is put on the pastern at night. This protects the elbow while it heals itself. Pressure bandages may help windgalls, as may the regular use of an embrocation or even a working blister to tighten the skin. Cold hosing combined with pressure stockings will generally help a big knee. A bursa may be drained and some of the fluid replaced with cortisone but this is a task for the vet. Any bursitis may prove unsightly and difficult to get rid of but is often harmless; however, at the poll or withers, problems such as poll evil and fistulous withers may develop.

Contracted Heels

Symptoms: The horse may go lame.
Causes: Excessive paring of the bars and heels and leaving the toe too long. Thrush (see below) reducing the frog is an alternative cause.
Treatment: Keep the toe short; this method takes several months. A pad or bar may also be used to induce extra frog pressure.

Dislocation

A partial dislocation is called *subluxation*. These are both very rare in the horse. The more common bone location problem is upward fixation of the patella.

Symptoms: The affected hind leg is stretched out and back.

Causes: A weakness in conformation may predispose towards this problem. It is more common in young horses and they may well grow out of it.

Treatment: The condition may right itself if the horse is made to step forwards or back or turn. In other cases manipulation of the patella and leg may be needed. It may be necessary for the horse to wear a higher-heeled shoe.

Exostoses

Symptoms: The growth of excess new bone, forming a bony enlargement, following a tear of the periosteum or a bruise. The horse may first go lame and there may be heat, but eventually, when the bony growth has settled, the horse will usually go sound. However, flexion of a joint may be reduced. X-rays (radiographs) may help diagnosis.

There are many examples of exostoses, e.g. high ringbone, on the front of a pastern joint; low ringbone, on the front of the coffin joint; and false ringbone, not on the joint (non-articular). Other examples are osselets on the front of the lower cannon and upper long pastern; splints on the splint bone; bone spavin at the front of the hock; and occult spavin on the articular surfaces of the hock.

Causes: There may first be inflammation of the bone (ostitis) or inflammation of the periosteum (the skin around the bone). This is called periostitis. There may be a fracture, and this is one possible cause of a splint. Exostoses of the sesamoid bones may follow inflammation (due either to strain at the ligament/bone junction or to fracture). This is known as sesamoiditis. Upright joint and too much roadwork may predispose the horse to various forms of exostoses of the lower leg. Some people consider that ringbone is hereditary.

Treatment: A mild blister is sometimes used. Phenylbutazone eases the pain, as in arthritis. De-nerving is carried out in extreme cases. Rest is often prescribed, but walking exercise has proved beneficial. Improved nutrition and physiotherapy may both help.

Fractures (A break in bone or cartilage)

Symptoms: There may be an incomplete break as in the 'greenstick' fracture of a youngster. The fracture may be simple, crushed (comminuted), compound or open, with the outer skin broken. It may be only a hairline fracture going a little way into the bone, as found, for example, in sore or bucked shins. The break may be heard and may be seen. The horse will be lame and in great pain.

Causes: The lower leg may break under stress during exercise; the cannon may break during jumping. The pelvis may fracture if the horse slips up on the road and lands heavily on its hip. Any bone may break from the trauma of a severe blow.

Treatment: Immediate immobility is essential. The decision must be made as to whether treatment is realistic or whether the animal should be put down. Poulticing will provide helpful heat for sore shins.

Laminitis (Founder)

Symptoms: Hot feet in which the laminae are inflamed. The horse stands leaning backwards with the hindlegs well engaged and the fore feet toes not taking weight. The animal is in pain and shows it. The foot may have rings of growth caused by previous attacks.

Causes: Rich pasture, too long feet and lack of work leading to overweight all combine to give fat ponies laminitis. Any horse or pony which breaks into the food store or is on fast-growing grass may get a sudden excess of starch or sugar, which creates toxin formation in the intestines; these poisons circulate and damage blood vessels, especially in the hoof. Similarly, a retained afterbirth or severe inflammation of the gut will both produce inflammatory toxins.

Treatment: Call the vet. Apply ice packs to the feet or cold-hose for five minutes and then make the animal walk to aid circulation while waiting for the vet. Reduce the diet, trim the feet and force exercise at least twice a day for a week. This is a case for close co-operation between owner, groom, vet and farrier.

Pony-type laminitis is avoided by allowing such animals limited grazing by tethering or leaving in a yard where they get straw, cut grass or hay plus one hour's grazing daily. Spring grass is the worst offender, but the extra moisture in late summer may produce a second flush of grass. Keep the animal's feet well trimmed.

If the wall of the foot separates from the sole at the toe, this is called seedy toe. The crumbling horn must be cut away and the cavity packed with antiseptic paste. Such feet will need rocker shoes. Chronic laminitis may lead to rotation of the pedal bone and a dropped sole.

Navicular Disease and Pedal Osteitis (Ostitis)

Symptoms: Shorter stride; slight lameness on the day after work; resting a front foot in the stable; stumbling; pivoting on the front feet when turned in its own length; heat; pain from pressure on the sole; lameness, reducing with work.

Causes: Bruised or punctured sole or the result of too much work on the hard with flat, thin-soled feet may all cause pedal osteitis. Navicular disease is caused by blood clots blocking the arteries serving parts of this bone, thus causing these parts to die.

Treatment: X-ray examination will help to show which of the two bones is involved. The farrier may fit shoes with rolled toes to reduce stumbling. He may fit wedges to raise the heels and lessen the angle of the deep flexor tendon over the navicular bone. The horse may be worked on pain-killing drugs as a short-term expedient. Blood-thinning drugs (such as 'Warfarin') or blood vessel dilatory drugs (such as isoxsuprine hydrochloride) may help circulation to such an extent that damaged areas may even be replaced by new bone. As a last resort, the foot may be de-nerved. This makes the horse legally 'unsound' and must be disclosed upon sale.

Abscess at the Coronet

Symptoms: A sore swelling above the coronet, caused by pus in the foot, breaking out at the top of the foot.

Causes: Gravel; foreign matter gaining access to the foot at the white line. Pricked sole (either a nail driven wrong by the farrier or a sharp object, which may still be embedded). Binding nail (a shoeing nail driven too close to the sensitive laminae). Quittor (damage to top of lateral cartilage). (A tread – another foot bruises the coronet – may appear similar.)

Treatment: Treat the cause of infection in the foot. If necessary, a hole may be cut in the sole by the vet or farrier to let the poison out. This may be assisted by poulticing. Keep the sore at the coronet open and free-draining to let pus out. Pack the holes with antispetic on cotton wool and cover with clean bandage. Always make sure that the animal is protected against tetanus.

Sand Cracks

Symptoms: A split running downwards, from the coronary band, in the wall of the hoof.

Causes: Treads (see below) or blows to the coronet.

Treatment: The farrier may put a groove across the hoof at the bottom of the crack to stop it running down. Alternatively, he may insert a clench, a special staple to hold the horn together. If the crack runs to the bottom of the foot, the farrier should seat out the shoe under the crack. Feed biotin and/or methionine (gelatine).

Self-inflicted Damage

This may take several forms, treads, over-reaches, speedy cuts and brushing.
Symptoms: Cuts or sore places caused by the horse's own feet, located as follows:
    treads – coronet, often caused by other horses
    over-reach – heels or back of tendon on front legs
    speedy cuts – inside cannon
    brushing – inside fetlock or coronet
Causes: Tired animals or those with poor conformation moving on poor going or in crowded conditions.
Treatment: Treat as for a wound. Corrective farriery as needed. (*Note*: prevention is better than cure. Protective clothing, such as over-reach boots, brushing boots and Yorkshire boots, may be used. Exercise bandages will also protect from brushing.)

Sidebones

Symptoms: Ossification of the lateral cartilages which can cause lameness while the bone is forming. Sidebones may be felt as hard areas in the bulb of the heel and forwards from that area.
Causes: The Horse Breeding Act 1918 lists sidebones as hereditary and an unsoundness. However, ossification of this cartilage is normal, and premature ossification with temporary lameness is probably due only to excessive roadwork.
Treatment: Temporary rest.

Sprains and Strains

These are synonymous and especially affect the front leg.
Symptoms: Pain giving rise to lameness; heat and swelling.
Causes: Sudden stress on a tendon or ligament beyond its normal
limits, causing torn fibres. Upright pastern or 'back at the knee'
may be a predisposing cause as may deterioration in the
conformation of the foot, or lameness from some other cause. The
toe landing on a stone or the heel hitting softer ground when the
horse is stressed and exhausted, e.g. the last run of the day when
hunting, may also cause strains and sprains.
Treatment: Immediate application of cold to control swelling and
of a pressure bandage over padding to give support. The horse
should be boxed home and given complete rest until the heat and
swelling subside. Poultices may prove useful at this stage. Walking
exercise, cold hosing and regular massage form the later stages of
treatment. Muscles may respond well to ultrasonic or similar
treatments. Firing, carbon implant and split-tendon operations all
have their supporters, but repair is slow and difficult. Time is the
greatest healer.

Thrush

Symptoms: A smelly frog with moisture in the cleft. There is
sometimes lamenesss.
Causes: Standing in dirty bedding and failure to pick out the feet
regularly.
Treatment: Keep both bedding and feet clean. Clean the feet using
a stiff brush and disinfectant solution. Trim the cleft of the frog to
allow access of air. Treat the area regularly with an antiseptic
dressing or Stockholm tar.

4 Systems of information and control

Information and control are provided by the nervous and sensory systems and the ductless glands. The nervous system governs reasoned and co-ordinated movement. It stores information, evaluates situations against experience and instinct, makes decisions, and also sends commands to the muscles. The sensory systems act as an information service about matters both inside and outside the body. The external senses are the ears (hearing), the eyes (sight), the nose (smell), the mouth (taste), and the skin (feel of heat, pressure and pain). The ductless glands are small centres that receive information and control some aspects of the body in a way quite different from the nervous system. Their efficiency is vital to the well-being of the horse, but as yet they are imperfectly understood.

A knowledge and understanding of these systems will help to answer many questions that horsemasters may ask themselves about the horse, such as 'Why does the horse do it?', 'What will it do next time?', 'How did it know?'.

The Nervous Systems

The horse reacts to changes in the world about it. Each change, be it of light, temperature or any other stimulus, produces a reaction in the horse. The change or stimulus is first received and then conducted to a central control system, which interprets the message and causes appropriate action to be taken. This is the central nervous system (C.N.S., see Fig. 4.1) and consists of the brain and the spinal cord.

Central Nervous System

The brain is placed for safety in the horse's skull. It receives all messages from the senses, via the sensory nerves, and puts this information together to form an understanding of its immediate environment. This information can be stored as experiences. Thus, the horse can learn and can associate a past experience with a present happening and thus anticipate the next event.

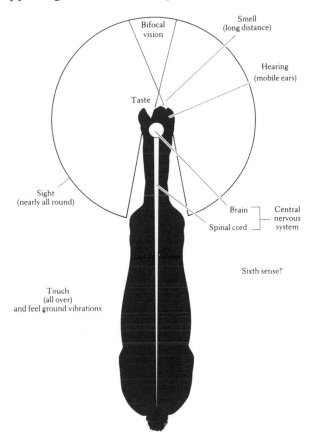

Fig. 4.1 Central nervous system and senses.

Although it has well-developed senses and an adequate memory, the horse lacks the powers of imagination or foresight. While the horse has an inborn instinct for self-preservation, it has no conception of death. It cannot cast its mind forward in tackling a situation. However, the horse may be influenced by its natural instincts and appetites.

Based on the horse's decision about its next action, the brain sends out messages to the body through cranial nerves in the skull and through the spinal cord which runs down the centre of the spinal column ('backbone') and on through spinal nerves that emerge between the vertebrae to go to all parts of the body.

Peripheral Nervous System

This includes the cranial and spinal nerves, plus the autonomic nervous system that controls the digestive system, much of the urinogenital system, the movement of blood by the heart and some of the activities of the glands. The peripheral nervous system also includes the motor and sensory nerves whose job is to transmit instructions to the muscles and to receive stimuli, respectively.

Aspects of Behaviour

Instinct

Instinct is inherent or innate ability. A foal will get up, feed, drink and walk without instruction. It also learns as it progresses. A foal moves unevenly when compared with an adult horse, but it acquires greater skill rapidly. When a woman in high-heeled shoes walks, her ankle makes small adjusting movements continuously; this is an acquired skill. A Thoroughbred horse has been bred to make similar compensating movements better than any other horse. One aspect of training horses is to teach them to move better and thus be more skilful in terms of movement than they are in the natural state.

The instincts guide the horse in its attempts to survive, to nourish itself and to reproduce.

Behaviour

Behaviour may be group or individual. Each horse is a distinct personality with a different temperament from its fellows. Temperament varies from horse to horse and can be observed not only in its actions but also in its facial expressions. Horses soon learn to recognise different situations and people. They can be nervous, happy, sensible, brave, cowardly or stupid – as can human beings!

As a group, horses tend to be nervous, responding quickly and violently to changes in environment. A horse is made this way because that is how it escapes from its enemies. It is not its nature to fight, except over food or sex. In a group, horses will establish a 'pecking order'. A horse (which need not be a male) will put others in their place, and within the group this order runs all down the line. Thus, when feeding hay in a field, it is essential to so place it that each horse can feed without being threatened by its fellows.

In the breeding situation, males will fight from about two years of age and upwards. Generally mares will not fight, but they will have only a limited period when they will receive a stallion.

Reflex Action

This is an automatic or unconscious response of a muscle or a gland to a stimulus. It is an immediate and involuntary response, such as coughing. Another example of a reflex action is the postural reflex, which allows a horse to sleep standing up. Its auditory reflex makes it turn its head towards a new sound. Happily for the horsemaster, the tonic neck reflex makes it less likely that a horse will kick or buck if its head is raised.

Inaction

Inaction may be caused by lack of either feeling or of the ability to move. Lack of feeling may be induced by anaesthetics, which are chemicals or drugs that prevent the passage of nerve impulses. A local anaesthetic is used to deaden a particular area, and a general anaesthetic affects the central nervous system, thus causing unconsciousness. If used in excess, a general anaesthetic will cause death by depressing the vital reflex centres in the brain.

Lack of the ability to move can be caused by paralysis, which in turn is caused by nerve damage. Damaged nerves may recover, but dead ones cannot do so. However, they may be replaced to some extent by other nerves taking over their duties if given sufficient time.

Sensory Systems

The senses are the five faculties of sight, hearing, smell, taste and touch. Through these senses the horse perceives the external world.

Sight

The eye (Fig. 4.2) is the organ of sight. An image enters through the lens and is focused on the retina at the back of the eyeball, just as in a camera the picture is focused on the film. In comparison with other animals, the muscles that change the shape of the lens are poorly developed in the case of the horse. This could create difficulties in focusing, but it is thought that the horse can compensate for this by viewing objects at distances on different parts of the retina.

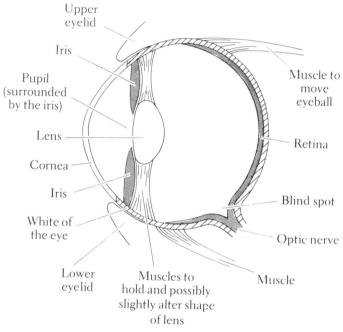

Fig. 4.2 The eye.

The distance from the lens to the retina varies very slightly from top to bottom. So, to help focus on close objects, the horse can raise its muzzle and form the picture at the top of the retina. To look at distant objects it can tuck in its muzzle and view the object at the bottom of the retina. Thus, when the horse is grazing, the grass (viewed down its nose) is in focus, as are distant objects viewed under its brow. Horses probably suffer from astigmatism so that part of the picture they view is a little blurred.

The arrangement of the horse's eyes (set at the side of its head) enables it to be on the watch for enemies coming from any direction. Within this wide panoramic view, however, the horse probably has a poor definition of distance. To get a better idea of an object, the horse will turn its head and focus both eyes on it, probably concentrating on the object to the virtual exclusion of the rest of the picture.

When the horse is jumping, it needs to alter the angle of its head to keep the obstacle in focus until about the moment of take-off, when it loses sight of the obstacle itself.

It is believed that the horse has poor colour definition. Horses may well see colours differently to humans. As it has no imagination, the horse probably cannot distinguish stationary objects far from it. Possibly, too, the horse adjusts more slowly than humans to changes in light intensity and so may hesitate before going willingly from light to shade.

Although it is hard to be certain about such matters, the horse can probably only see about 150 m (around 500 ft) and only achieves recognition at about 60 m (200 ft). The horse apparently looking into the distance may well be receiving information through other senses.

Hearing

The ear (Fig. 4.3) is the organ of hearing. The ears of a healthy horse are always on the move. Horses have a keen sense of hearing – far better than that of humans – and by turning their ears towards a sound they can pinpoint accurately the direction from which it is coming. This is an evolved protective mechanism.

The horse can convey a certain amount of meaning with its voice, from the gentle whickers and whinnies of a mare to her foal, through the neigh of a horse turned into a strange paddock, to the wild shriek of an angry stallion. The horse also expresses meaning by snorting. The interpretation of sound is aided by behaviour and situation. By constant repetition, the horse can be taught to understand a short range of human verbal commands.

The ear is divided into three parts. The external ear is the mobile part which turns to catch sounds and can also denote mood. It continues into the head as far as the ear drum (tympanic membrane). The middle ear is a cavity behind the ear drum, and to maintain an equal pressure with the outside air it connects to the back of the throat (pharynx) by the Eustachian tubes, one on each side. The middle ear contains the three little bones that have the same 'horsey' names in all animals. They are the hammer (malleus), anvil (incus) and stirrup (stapes). These bones provide

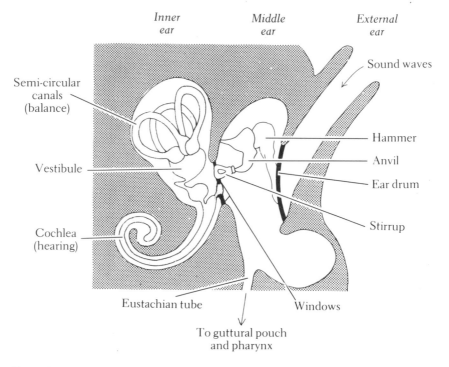

Fig. 4.3 The ear.

direct mechanical linkage from the ear drum to the oval window of the inner ear, the last of the three parts. This is tucked into a cavity within the skull. It has two parts: the cochlea receives the sound as fluid movement detected by hairs; the other part consists of a vestibule and three semi-circular canals, which are concerned with the balance of the horse. One of these is longitudinal, one transverse and the third at right angles to give three-dimensional balance.

Smell

The nose is the organ of smell. A horse breathes in only through its nose and not through its mouth. Smell is a very important sense to the horse. In feeding, the horse blows into its feed bowl to test the smell and it may refuse food which has an unfamiliar smell. Stallions can detect an 'in-season' mare by smell at 200 m (600 ft).

Some people blow up a horse's nostrils to communicate friendship, just as horses do to each other. Interesting-smelling compounds may also entice a horse, e.g. ginger.

The horse has nostrils which it can dilate. These lead into the nasal chamber, which is long and contains coiled bones covered with moist membranes well supplied with blood. The olfactory or smell nerve cells are sited in these mucous membranes, which have tiny hairs over their surface. It is these hairs that actually detect the smell as it goes into solution on·the moisture of the membranes.

Taste

The mouth contains the organs of taste, which consist of taste buds. These are little groups of cells which are found at the end of each taste nerve fibre. They are found mostly on the tongue, but also on the palate and in the throat. The information about taste is fed to the brain as degrees of salt/sweet/bitter/sour. This taste information is received with information on smell and texture to give a composite impression.

Horses like saltiness and sweetness but they dislike bitterness or sourness. It is because of this that most worm powders are sugar-based, and some people rub salt on their hands before examining a horse's mouth or teeth.

Preparations designed to stop horses chewing wood are bitter tasting, but horses like to chew fence rails cured with preservative salts. Usually, horses are very fussy feeders and will reject unaccustomed tastes. Similarly, they will generally reject fats and meats, but on occasions will eat such things as acorns and yew leaves, which are bitter tasting to humans and may in fact be poisonous.

Feel

The horse's skin has specialist nerve endings to receive feel, which may be divided into the five sensations of touch, pressure, cold, heat and pain. These nerve endings are limited in number in most areas, and therefore an injection can miss a nerve by chance and be completely painless. On the other hand, nerve endings are densely grouped in the muzzle. Some sensations of feel – such as gut pain in colic – can come from internal parts of the body.

There is another sense like feel, which is that of 'awareness' of the body. This muscle sense or proprioception enables the horse to know what its limbs are doing and how they are responding to the muscle contractions that control movement.

A third sensation of a similar nature is the organic group. This group of sensations indicates conditions of bodily need, such as hunger, thirst, need to micturate (relieve the bladder) or copulate.

A further sensation of feel is the horse's awareness of ground vibrations. Thus, often before it can see or hear someone approaching, the horse will show its awareness of the approach.

The horse is receptive to rubbing, scratching and nibbling on certain areas. A pony that refuses to be caught and turns its quarters towards the would-be catcher may be caught by scratching its rump, though one should watch the ears for signs of intent to kick. The head-shy horse may be relaxed by scratching its withers and the crest of its neck prior to touching the head.

Sensitive areas can be dulled by misuse. The bars of the mouth become insensitive with hard hands and sharp bits. The horse's sides become dulled with constant kicking.

Sixth Sense

Horses can detect certain things that human beings cannot. For example, horses will not go near radioactive material and also seem to be aware of approaching weather. Some people go so far as to suggest that horses are sensitive to 'psychic vibrations'. Horses certainly seem to be aware of the state of mind of humans with whom they have contact. Just as lie detectors attached to a person's skin can detect changes in thought patterns, so some animals appear to receive similar signals. Possibly it is for this reason that certain people have 'a way with animals'.

Although horses have no imagination, experience soon teaches them to associate. Thus a stranger smelling of surgical spirit tends to mean a sharp prick from a hypodermic needle! Being plaited-up means going hunting or to a show. This is so exciting to some horses that they will not eat breakfast. Thus, not everything to which we attribute sixth sense is beyond comprehension, but there are some things about the horse's awareness of its environment which are not yet understood.

Endocrine System

The endocrine system (ductless glands; see Fig. 4.4) controls the horse's patterns of behaviour. The system consists of a series of glands, which secrete hormones directly into the blood or lymph streams. A gland is an organ secreting chemical substances for use

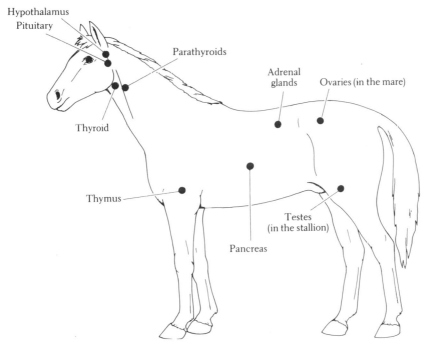

Fig. 4.4 Ductless (endocrine) glands.

in the body. Those with which we are here concerned are called endocrine glands, in contrast to exocrine glands which secrete through ducts to the outside or to the digestive tract. The endocrine glands produce hormones, which travel in the blood or lymph streams to parts of the body often far removed from the gland itself and which exert a specific effect on certain tissues. The word 'hormone' comes from the Greek *hormon*, which means 'to stir up' or 'to arouse activity'. In some cases, the presence of a hormone in the blood system produces a concerted effect on the body, creating a specific state or condition. These glands control conception, gestation, parturition, metabolism, growth, puberty, ageing, aggression, passion, and so on. The usual classification of the endocrine glands is as follows.

Hypothalamus

The hypothalamus is a nerve control centre at the base of the brain, and is concerned with hunger, thirst and other autonomic functions. It also releases factors that control the all-important pituitary gland.

Pituitary

This is attached by a stalk to the base of the brain and has been called 'the master endocrine gland'. It is the leader of the endocrine orchestra. The pituitary gland consists of two lobes, the anterior and the posterior. The anterior lobe produces follicle stimulating hormone (F.S.H.), and luteinising hormone (L.H.), which both act on the ovaries in the female. It also affects the testes of the male, and the thyroid and adrenal glands of both sexes. The posterior lobe produces hormones affecting the kidney, the uterus and the mammary glands.

Thyroid

This gland is situated on either side of the larynx and produces hormones composed largely of iodine. A shortage of this chemical thus affects the operation of the thyroid. The thyroid gland controls metabolism and growth. An underactive thyroid results in a lack of energy and a tendency to overweight. If this gland is overactive, the reverse occurs. The thyroid gland has been likened to a blacksmith's bellows governing the fires of life. In normal function it has an automatic self-righting mechanism so that any excess hormone will act to shut off the pituitary stimulus.

Parathyroids

These are a group of four small glands situated near the thyroid and control calcium and phosphorus in the body.

Adrenal Glands

These are located close to the kidneys and produce two hormones, cortisol and adrenalin. Cortisol has a variety of effects, including the control of inflammation. Adrenalin is a stimulant in response to stress: it increases the heart rate and blood pressure. It prepares the body for fight or flight.

Pancreas

The pancreas is situated behind the stomach and secretes insulin, which controls the level of blood sugar. This gland also produces digestive juices.

Thymus

The thymus is just under the breastbone between the lungs. 'Sweetbreads' as sold by the butcher are the thymus glands of calves. This gland is particularly large and active in foals because it is concerned with immunity. It is also a source of blood lymphocytes.

Other Ductless Glands

In the female, the uterus acts as a ductless gland to produce prostaglandin, which affects the body in many ways, particularly in bringing the mare into season (see chapter 14). The ovaries, in addition to producing eggs, secrete oestrogen and progesterone. Oestrogen is responsible for the mare's behaviour and changes in her sex cycle, and progesterone, which is also produced in the placenta and adrenal glands, also has an important part to play in the mare's cycle (chapter 9). The testes produce the male sex hormone, testosterone, which is responsible for the male characteristics and development.

DISORDERS OF THE SYSTEMS OF INFORMATION AND CONTROL

This section concentrates on those disorders and malfunctions commonly met in practice. In fact, the senses generally give little trouble, with the exception of the eye, where the commonest problem is injury. The nervous system also generally functions normally throughout the life of the horse, but occasional nervous diseases and conditions may be encountered. There are also nervous habits which are regarded as vices and which cause considerable concern to the horsemaster. Glandular disorders are comparatively rare.

Ailments of the Eye

Bruising

Symptoms: Swollen eyelids, but without damage to the eyeball.
Causes: A blow is the most common cause, e.g. from a branch when riding through a wood.

Treatment: Hot fomentation, using a teaspoon of kitchen salt in half a litre of boiled, cooled water. Any lacerations to the eyelids must be stitched at once by the vet.

Cataract

Symptoms: Jumpiness, because the horse has areas of partial blindness within his total field of vision. The condition can only be diagnosed through the use of an ophthalmoscope, which reveals an area of opacity in the lens.
Causes: It may be present at birth or may be caused by an injury or infection.
Treatment: A small cataract found on veterinary examination will not generally affect the horse. The condition may deteriorate.

Conjunctivitis

Symptoms: Inflammation of the membrane inside the eyelid: the eye itself may look bloodshot. The eyelids will be swollen and there may be tears and a mucus discharge from the eye.
Causes: Injury or a foreign body, and also irritation, cold, allergy or infection. Also associated with blocked tear ducts.
Treatment: Remove any foreign body present. Wash with warm water which has been boiled and allowed to cool. Flush with a cool mixture of a teaspoon of boracic acid or Epsom salts to 0.5 litre (about a pint) of boiled water several times a day. If discharge continues, an ophthalmic ointment from the vet is needed.

Entropion

Symptoms: An ingrowing eyelid. Irritation.
Cause: A foal may be born with this condition.
Treatment: It is essential that it be spotted and treated at a very early stage by the vet, who will probably stitch back the ingrowing eyelid for a short time, thus effecting a cure.

Keratitis

Symptoms: Inflammation of the cornea (the eye's transparent front covering) with tightly closed eyelids.
Causes: A blow, a foreign body, turned-in eyelids or an infection.
Treatment: Call the vet.

Periodic ophthalmia (Moon blindness)

Symptoms: Generally, only one eyeball is affected. The eye

undergoes progressive inflammatory changes and after a few weeks there is a tendency to partial recovery. The symptoms later recur.
Cause: The cause seems to be unknown but the ailment is possibly bacterial or viral in origin.
Treatment: Keep the horse stabled in relatively dark conditions and consult the vet.

Photosensitisation

Symptoms: Head shaking.
Cause: Bright sunlight.
Treatment: Discuss this condition with the vet. However, as he or she will advise, this is not the only possible reason for head shaking.

Nervous Disorders

Concussion

Symptoms: Loss of or reduced consciousness; dilated pupils; laboured and irregular breathing.
Cause: A severe bang on the head.
Treatment: Give the horse space and quiet. The head and spine may be cooled by sponging with cold water. Send for the vet.

Shivering

Symptoms: Involuntary and spasmodic muscular contractions, usually of the hind leg and without pain.
Cause: Although this may follow serious illness or a bad fall, it is a progressive nervous disease.
Treatment: None known.

Staggers

Symptoms: Loss of equilibrium.
Causes: An infection of the brain, grass sickness, wobbler syndrome, or possibly poisoning.
Treatment: The vet must be consulted.

Stringhalt

Symptoms: An upward jerking of one or both hind legs due to the excessive flexion of the hock. It may not occur on every step.
Cause: Not known but it is listed as a hereditary disease by the

Horse Breeding Act 1918.

Treatment: Nothing need be done about it, but eventually the condition will get worse. However, for many years the horse can be used as normal and its effectiveness in galloping and jumping is not impaired. Nevertheless, it must be declared as an unsoundness. There is an operation for this problem.

Nervous Habits (Stable Vices)

These may be caused by boredom or idleness, are difficult to cure, and a horse with such a habit is classified as unsound.

Crib-biting

Symptoms: The upper teeth may be unnaturally worn and on discreet observation the horse will be seen to grab hold of the manger (the crib) or the top of the door or anything handy, drawing air in through the mouth and swallowing it down to its stomach. This causes indigestion and unthriftiness.
Cause: Not known (neither copying another horse nor boredom will explain many cases). Chewing rails in the paddock or stable woodwork should be discouraged lest they lead to this vice.
Treatment: Turn out by day. Give regular work. Remove anything movable on which he can crib, and creosote all woodwork. Cover any remaining edges with a specially-made preparation.

Weaving

Symptoms: Swinging the head and neck from side to side, especially when placed in a strange stable. In the advanced form, the horse will rock from one fore foot to the other.
Causes: The horse may catch the habit by watching other horses. Boredom, nervous tension.
Treatment: Turn out by day and give regular exercise. A full grid may be fitted in the top door, but some horses will weave behind this. An antiweaving grid is probably the best solution and may effect a cure. Alternative treatments include hanging a brick on a string in the centre of the door or tying up for the afternoon. Offering a little fresh-cut gorse is a useful distraction, but an unpleasant chore in the gathering.

Wind-sucking

Symptoms: The horse arches his neck and swallows air without catching hold of anything. Some horses will only do this when left alone, and so it may be hard to observe. Like crib-biting, it may cause indigestion and unthriftiness.

Cause: May develop out of crib-biting.

Treatment: Fit a 'cribbing strap', which buckles around the top of the neck and catches the horse in the throat if it tries to arch its neck. Turn out by day and work regularly.

5 Circulatory system

The body is a complex structure consisting of many cells, each with requirements which must receive attention on demand. Within the body, therefore, there must be a transport system which functions with unfailing efficiency.

The body's transport department is based on the heart. Essentials are carried to the point where they are needed, defence forces move along the system looking out for trouble and ready to summon help when necessary. Heat is evenly distributed throughout the body to meet the animal's needs. Waste products are collected and carried to disposal points.

Tasks of the System

The circulatory system carries the following essentials:

(1) *Oxygen* from the lungs to all of the body cells, especially muscles.
(2) *Carbon dioxide* from the body cells especially muscles, to the lungs.
(3) *Water* and *nutrients* from the gut to the body cells.
(4) *Waste* from body tissues to the kidneys.
(5) *Messages* (hormones) from the endocrine glands to other organs.
(6) *Defence forces* to sites of attack.
(7) *Heat* from the centre of the body to its surface, as required.

Additionally, the system bathes the body cells in a uniform environment, and it resists leakage by virtue of the blood's ability to clot.

Blood

The constituents of the blood can be summarised as follows:

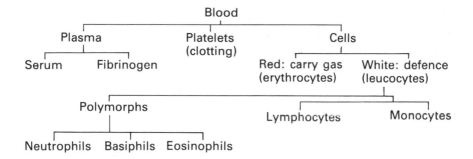

When a horse is off colour with no clear symptoms of disease, the vet may take a blood sample and produce a report showing the analysis of the blood. The vet will put some comment on the report, but it aids understanding to have some knowledge of blood itself.

Blood consists of a fluid called plasma, in which there are red and white blood cells or corpuscles of which red are the more numerous and give blood its characteristic colour. Blood is normally fluid during life and when freshly drawn from the body, but it clots or solidifies rapidly on exposure to air.

Plasma

Plasma contains fibrinogen (which aids the clotting of blood), together with platelets. The remainder of the plasma is called serum and this contains water, proteins, glucose, lipids, amino acids, salts, enzymes, hormones, antigens, antibodies and urea. It is this fluid which bathes the cells of the body. Plasma consists of about 90% water and its content is influenced principally by food and water from the gut, the requirements of the body, and the action of the kidneys.

Blood Cells

The study of blood cells is called haematology. Blood cells are either red or white. Red blood cells originate in the bone marrow; they are also known as erythrocytes, and have the power of absorbing oxygen. White blood cells (leucocytes) defend against disease by attacking and destroying harmful germs, and are subdivided into two broad categories, the granulocytes (or polymorphs) and the agranulocytes, which are again subdivided into lymphocytes and monocytes, both of which originate in the lymphatic system (as discussed later). In contrast polymorphs originate in the bone marrow and are subdivided into neutrophils, which engulf bacteria to form pus; basiphils, which help control inflammation; and eosinophils, which detoxify foreign proteins.

The red blood cells contain haemoglobin. This substance has the ability to combine with oxygen and carry it as oxyhaemoglobin from the lungs to the muscles. Haemoglobin also carries carbon dioxide. Thse two functions are essential for the efficient working of muscles.

The packed cell volume (P.C.V. or haematocrit) refers to the percentage of whole blood constituted by red blood cells.

Analysis

A blood sample will indicate the horse's state of health. For example, if the red blood cell count or haemoglobin level is low, the horse is anaemic and must be treated accordingly. A high percentage of lymphocytes and monocytes may indicate some chronic disease. An excess of eosinophils suggests that there is a high percentage of invaders, such as worm larvae, in the blood.

The study of the horse's blood as a guide to performance and treatment is complicated by the ability of the horse to mobilise reserves very quickly; the spleen of the horse acts as a reservoir of blood cells. Thus, a different picture in some aspects (particularly the haematocrit) can be given by the same horse at different times on the same day, depending on alterations in activity and environment, particularly a stimulating change such as a journey, or even a vet who is a stranger.

A blood sample can still provide useful clues which, considered as part of the overall picture, give important pointers to health improvement. Anaemia, larval infestation, virus attack and azoturia are all examples of cases in which the blood can supply necessary information to aid diagnosis and treatment.

The Heart

The heart is a hollow and cone-shaped organ consisting of muscle contained in a protective cover (the pericardium). It is sited in the centre of the chest. The heart is divided into four internal compartments. The upper chambers are called atria (formerly auricles) and the lower are known as ventricles. The function of the heart is to keep up the circulation of blood.

The vena cava brings deoxygenated blood back to the right side of the heart. The blood is collected in the top chamber (atrium), and goes through a non-return valve into the lower chamber, which squeezes the blood up the pulmonary artery to the lungs.

The pulmonary vein brings oxygenated blood back from the lungs into the left atrium; it then goes down through a non-return valve into the left ventricle, which squeezes and forces the blood along the main artery of the body (the aorta) under pressure.

Both sides of the heart operate in parallel, and so both atria fill and then contract, sending the blood into the ventricles. In turn, both ventricles contract and send the blood up the pulmonary artery and the aorta. When a heart chamber contracts, valves close to stop the blood going back the way it came, and as that chamber relaxes these valves reopen. The sound of the contractions and the operation of the valves produces the heartbeat, which can be heard when the horse is stressed or, with the aid of a stethoscope, behind the left elbow. The sound is written '*LUBB*-dup – *LUBB*-dup – etc.'.

The normal heartbeat at rest is around 35 to 45 beats a minute but at top speed it can rise to about 200. If the rate rises and stays up when the horse is at rest, this is a sign of distress. A slight irregularity of heartbeat is not unusual in horses and may give no cause for concern: only the vet can say. A 'murmur' is the sound of a heart valve working imperfectly, but again this may not be a cause for concern – though it may constitute legal unsoundness.

Circulation

The circulatory system is shown diagrammatically in Fig. 5.1. Oxygenated blood from the heart is carried round the body under pressure in muscular thick-walled tubes called arteries, which gradually diminish in calibre as their length increases. The main artery leaving the heart is the aorta and its first branches supply blood to the heart itself. It has further branches off to all parts of the body, supplying every organ and structure. The next major branch (the brachiocephalic) goes to the head and forelegs. The aorta goes through the diaphragm and gives off a large branch called the coeliac artery, which supplies the stomach, liver and

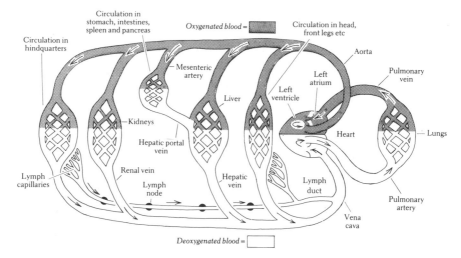

Fig. 5.1 Circulatory system.

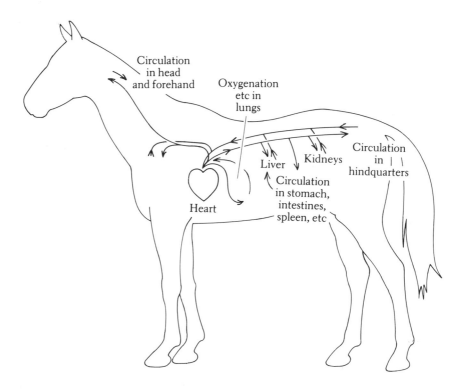

Fig. 5.2 Circulation in the horse.

spleen. The intestines are supplied with blood by the mesenteric arteries: the large cranial at the front, and the smaller caudal at the rear. The renal artery supplies blood to the kidneys, and the iliac arteries supply the area of the hindquarters. (See Fig. 5.2.)

Every artery divides into smaller arterioles and these subdivide into the delicate capillaries which permeate throughout the body. Water, oxygen and nutrients filter out from the capillaries to the individual cells and some water returns in the same way, together with waste products. Capillaries unite to form venules and these in turn unite to form veins.

Venules are in effect small branches of a vein which receive oxygen-depleted blood from the capillaries and return it to the heart via the venous system. Veins are thin-walled tubular vessels that convey deoxygenated blood on its return journey to the heart. Veins are mostly named after their opposing arteries: thus, the vein from the kidneys is called the renal vein. The veins flow into the vena cava which takes the blood back to the heart.

The main exception to this general arrangement is that blood from the intestines needs to be filtered by the liver before going into general circulation. Such blood is collected into the hepatic portal vein, which divides within the liver into a capillary network for filtration purposes before regrouping to form the hepatic vein, which goes into the vena cava.

The whole system is called the systemic circulation system. Pulmonary circulation is the system supplying blood to the lungs. Deoxygenated blood is carried to the lungs by the pulmonary artery, and the pulmonary vein returns it in oxygenated form.

Lymphatic System

Lymph is an almost colourless fluid, chiefly containing white blood cells, which surrounds the tissues of the body. It comes from the blood and has to be returned to it. Some of this fluid is picked up by capillaries, but this collection system is not sufficient. It is paralleled by a second means of collection, the lymphatic system, which is both extensive and important.

Lymphatics are very thin-walled vessels, and their function is to collect the goodness from digested food and to prevent accumulation of fluid in any part of the body. The lymph capillaries flow into the lymph vessels, which have valves to ensure that the direction of flow is towards the heart. The vessels also have nodes or filters (which are bean-shaped masses of tissue) to check the lymph for infection and if necessary to produce lymphocytes and antibodies to cope with it. If there is infection in an area, the lymph nodes or glands will be enlarged and prominent. This is particularly noticeable under the jaw and inside the thigh.

Disorders of the Circulatory System

Anaemia

Symptoms: Pale gums, eyelid linings or other mucous membranes. Heart having to work harder to get oxygen around the body; it beats faster than normal. Lack of best performance and reduction in stamina.

Causes: Shortage of red blood cells or haemoglobin through haemorrhage (bleeding), infection, redworms, bots, or dietary deficiency.

Treatment: The vet must deal with any infection. The horse will need to work less hard during the recovery period. Folic acid, B_{12} or iron supplements may be required in the diet.

Prevention: Control redworms and bots. Grow deep-rooting herbs in the paddock or feed Russian comfrey or mineral supplement.

Azoturia (Set-fast or Tying-up)

Symptoms: Typically, the horse is excited, has done some work, has to wait and is unable to go foward. Examples include after the first draw out hunting, or between Phases A and B in a three-day event, or after warm-up waiting to jump at a show. However, it may occur after the horse is back home from exercise. The horse will sweat up and be unable to walk forward or will do so with difficulty. The muscles behind the saddle and those of the hindquarters feel hard. The horse is distressed by pain. The hind legs may be paralysed. Dark red, strange smelling urine may be passed. Severe cases collapse completely.

Causes: A disease attributed to management but some horses seem to be predisposed towards it. High corn diet and lack of exercise. Glycogen in the muscles is converted during exercise into lactic acid. In cases of azoturia, the lactic acid is not removed quickly enough and accumulates in the muscles, causing them to seize up. It is very like an attack of cramp.

Treatment: Do not move the horse. Rug the horse up and keep it out of draughts. If out, have the horse transported home. Call the vet, who may give drugs for the pain and inflammation. Hot fomentations may ease discomfort.

Prevention: Keep exercise ahead of food – as for lymphangitis (see below). A vitamin E diet supplement may help. The vet may prescribe an antacid or a buffer mixture in the diet.

Dehydration

Symptoms: A fold of skin pinched up will be slow to flatten out. After exercise, the horse is slow to recover its normal pulse and respiration rate. It is dull and lethargic. Prolonged stress such as a long journey or an endurance ride in hot weather can produce an audible diaphragm spasm ('the thumps'). This is a serious condition, the horse being near collapse, and it needs urgent veterinary attention.

Causes: Can be shortage of water, or the horse not drinking enough, or loss of fluid through scouring or excessive sweating from exertion, travel in a confined space, or fever.

Treatment: Ensure that the horse normally has access to water at all times. Check that the water is pleasant for the horse. Salt in feed may encourage it to drink more. The cause of any scouring or looseness in the droppings must be treated.

Dehydration symptoms after exercise require less than a half bucket of water, preferably containing a large pinch of salt and with glucose added. This is repeated at 10-minute intervals until the thirst is satisfied. Electrolytes may also be offered as a drink in water, but in severe cases may have to be given intravenously by the vet.

Endocarditis and Pericarditis

Symptoms: Often vague. Loss of interest in work and falling-away in condition.

Causes: The membranes lining the heart or surrounding it are inflamed by bacterial or viral attack.

Treatment: Cure the infection under the vet's direction. Months rather than weeks are required for convalescence, and rest is important.

Grass Sickness

Symptoms: Inflamed membranes. Difficulty in swallowing. Depression. No bowel activity. Dribbling. Mucus discharge from nostrils. Foul-smelling breath.

Cause: Unknown, but probably viral.

Treatment: None known. The disease is almost always fatal.

Haematoma

Symptoms: Lumps or bulges under the skin.
Causes: Haemorrhage (bleeding), either below the skin or into a muscle to form a lump, generally resulting from a blow. Can be large or small.
Treatment: Small haematomae are absorbed, but large ones require veterinary treatment and a few days' rest.

Lymphangitis (Big Leg)

Symptoms: Typically, the horse has a swollen hind leg which is warm and tender. The swelling often goes up to a line about the stifle. The horse shows symptoms of pain and has a raised temperature. Inflamed lymph vessels and nodes can occur in other areas.
Causes: An imbalance between a high corn diet and lack of exercise. It is often referred to (as is azoturia) as a 'Monday morning disease'. It may also be caused by infection.
Treatment: The vet will prescribe suitable drugs. The horse requires a comfortable stable, a laxative diet and hot fomentations applied several times daily to the limb. Lameness will limit exercise, but a little gentle walking in hand will help after the initial swelling has reduced slightly.
Prevention: Keep exercise ahead of food. If the horse is to have a day off, e.g. a Sunday after hunting on Saturday, cut back the food the night before. If possible, put the horse out in the paddock for an hour or more on the rest day.

Overheating

Symptoms: Horse appears exhausted, dull and lethargic, and is sweating. It may be trembling, with a raised pulse and respiration. Temperature may be as high as 42°C (108°F).
Cause: Prolonged or very strenuous exertion in conditions of high heat and humidity.
Treatment: Sluice the horse down with cold water. Wet towels over the head may help. Keep in the shade and in gently moving air to aid evaporation. Treat symptoms of dehydration.
Prevention: When high humidity combines with dry temperature, take great care to avoid dehydration. Avoid high speed plus long duration and cool the body with cold water as often as possible.

Strangles

Symptoms: Swollen glands under jaw. Nasal discharge. Temperature. Cough. Swallowing is painful.

Cause: Contagious specific bacteria.

Treatment: Isolate. Check spread of disease to other horses. Get the vet quickly as it is highly contagious. Rest. Sloppy food.

Transit Tetany

Symptoms: Distress, sweating, fast breathing, stumbling gait, all after a journey (especially of lactating mare in warm weather).

Cause: Lowering of the blood calcium level.

Treatment: Cool, quiet box, water to drink. Call the vet.

Tetanus (Lockjaw)

Symptoms: Increasing stiffness of movement. The horse stands in a stretched out manner. Tapping under the chin will cause the third eyelid to react. Later, the whole of the horse reacts to noise and touch. Muscular spasms. The jaws clamped together.

Cause: A bacterium found in the soil (*Clostridium tetani*). If this gets into an airless situation in a wound, toxins are produced which cause muscle spasm. Entry may be through a puncture wound or even a scratch.

Treatment: Call the vet. Prognosis is not good, but early diagnosis will help. Quiet, dark stable.

Prevention: Vaccinate with tetanus toxoid and give biennial booster injections. Very often this is combined with protection against equine flu. Pregnant mares may get their tetanus booster in the last month of pregnancy to help extend cover to the foal. The foal will need its own protection after about three months.

Thrombosis and Embolism

Symptoms: Depends on the site of the problem. The commonest site is the gut and the symptom is colic.

Causes: Obstruction of blood vessels by an attached stationary blood clot (thrombus). The commonest cause of these clots is redworm larvae in the system. If the blood clot breaks off and flows along the vessel and then jams across a small blood vessel, this is an embolism, and the area served by that vessel will suffer. An embolism may cause part of the body to go out of function for a short while. Sometimes the blood will sort out an alternative route around the system. The larval attack may cause a collapse of the artery wall (an aneurism).

Treatment: Treat the symptom (colic). Call the vet. Control worms, particularly larval stages migrating through the body.

6 Respiratory system

The main task of this system is to get oxygen into the blood. Without oxygen, all heat production and activity will cease. If the respiratory system falters, the horse will die within a few minutes. Of the horse's three major requirements, food, water and oxygen, the latter is crucial in the shortest time. A horse can go several days without water; it can go weeks without food; but death will occur if it is deprived of oxygen for a matter of minutes.

Tasks of the System

Like several other systems, the respiratory system has a primary function and a number of subsidiary but important ones. The provision of oxygen is the major function of the system. Its other tasks are:

(1) to remove carbon dioxide from the blood;
(2) to help temperature control by breathing out warm air and taking in cool air;
(3) to eliminate water: this is seen readily on a cold day, but happens at all times;
(4) to communicate by sound (voice production);
(5) to act as a sensory input: this is done through both smell and touch (via the nostril hairs).

Air Passages and the Lungs

Within the Head

Oxygen is taken down into the lungs along a highly specialised route. The extremity of this route consists of the nostrils, which are large, soft, gentle and inquisitive. They change in shape according to the horse's needs. The horse draws air in only through the nostrils and not the mouth. They are easily dilated and form part of the horse's facial expression when it is inquisitive or angry. Facial expressions are backed up by air blown from the nostrils and expelled as a blow or a snort. The hairs between and below the nostrils combine with the animal's sense of smell to investigate strange objects at close range. The airways in the head are shown in Fig. 6.1.

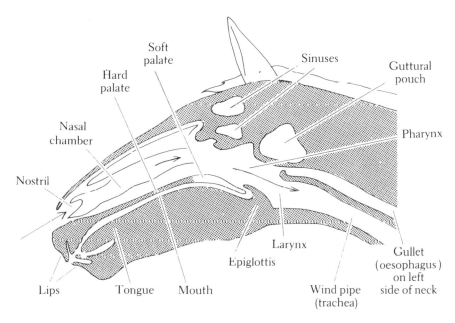

Fig. 6.1 Airways in the head.

The nasal cavities – one for each nostril – are divided from each other by a piece of cartilage. They are separated from the mouth by the hard palate and, higher, by the soft palate. The cavity is partially filled wth wafer-thin, curling bones, called the turbinate bones, which are designed to have a large surface area. This, like the rest of the cavity is covered by mucous membrane which helps to warm incoming air so that it does not strike too cold on the lungs. In warming the air in this way, the body loses heat when the air is expelled. This membrane, in the higher part of the cavity, contains the olfactory nerve endings, which detect smells.

In the skull at the front are air-filled cavities called sinuses, connecting with the nasal cavity. The maxillary sinus is above the molar teeth, and the others are called the frontal, the spheno-palatine and the ethmoidal sinuses. All exist in pairs, one on each side, and they give the skull strength and form, without excessive weight.

The pharynx or throat is the common passage for food and air, each coming from a different place and going down a different tube, with a crossover mechanism which shoots the food to be swallowed up and over the windpipe and into the gullet. When stomach-tubing a horse, it is necessary to pass the tube up the nostril because of the arrangement of the pharynx. Care must be taken to ensure that the tube goes down the gullet and not down the windpipe.

The Eustachian tubes come into the top of the pharynx. They allow air to pass to the middle ear, and connected to them are the guttural pouches which are sited just above the pharynx.

Within the Neck

Air passes from the pharynx through the larynx, which is the organ producing the voice. It is a box of jointed cartilage and the surrounding muscles can alter its calibre. The function of the larynx is to control the air going in and out, monitoring it for foreign objects which must be rejected. It is sited in the throat between the branches of the lower jaw, where it may be readily felt. Food and water pass over its opening (the glottis) on the way down the gullet, and it has a lid (the epiglottis) which closes automatically when the horse swallows. Anything wrong with the glottis or epiglottis causes coughing.

These organs and the vocal cords (which are thick and elastic) produce the horse's sounds of communication or voice. The larynx is thus commonly called the 'voice-box'. The horse can squeal, nicker, whinny and groan. The muscles that retract or draw back these cartilages are controlled by a branch of the vagus nerve which, curiously, is much longer on the left than on the right side. This nerve sometimes malfunctions so as to give wind problems, which are often confined to the left side of the larynx. Despite the larynx's function as a filter, some dust is inhaled by the horse, particularly when eating dry hay or when straw is shaken up when bedding down.

The windpipe (trachea) runs from the larynx to the lungs. It runs along the lower border of the neck and can easily be felt as far down as the entrance to the chest. It is a tube reinforced with rings of cartilage with overlapping ends, and is lined with microscopic hairs.

Within the Chest

The windpipe divides into two bronchi at the entrance to the chest, one branch going to each lung. From this point they divide and subdivide into bronchioles, which end as alveolar sacs. These in turn are subdivided into lots of little alveoli like grapes in a bunch, so as to give maximum surface area. (See Fig. 6.2.)

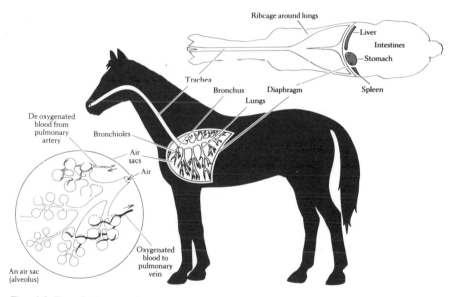

Fig. 6.2 Respiratory system.

The lungs are two large elastic organs. The horse's right lung has an extra lobe, but the shape of the lungs is not important as they completely fill the cavity of the chest (except for the cavity occupied by the heart) below the backbone and enclosed by the ribs and the diaphragm. The lungs are surrounded by a covering (the pleura), which is a smooth and slippery membrane that prevents friction.

The blood vessels in the lungs are so finely branched that the air is nearly in direct contact with the blood. Oxygen is able to diffuse through the fine layer of cells from the air into the blood, and carbon dioxide and water move in the opposite direction. The air is then exhaled back along the same route.

The diaphragm (or midriff) separates the chest from the belly. It is a strong and thin sheet of muscle attached to the inner sides of the ribs. It begins just in front of the loins, high under the backbone, sloping downwards and forwards to the breast bone. The major blood vessels and gullet go through it as they come from the chest, and like the lungs (which it touches) it is covered with pleura.

Breathing

Breathing is taking the air down into the lungs so that gaseous exchange with the blood is possible. It is one aspect of respiration. The second aspect is the activity in the body tissues where the oxygen is used.

The horse draws in air through its nostrils. The air passes through the larynx, down the windpipe and into the lungs. There, air and blood exchange the required elements and, after a slight pause, the used air is expelled. The process is repeated.

Air is drawn into the lungs by muscular expansion of the thorax or rib cage, and is expelled by elastic recoil of the rib cage. In the horse the rib cage is a single cavity whereas in many animals it is divided into two, with a separate lung in each. It is surrounded by ribs and separated from the abdomen by the diaphragm, which is dome shaped, with the top of the dome nearest the front of the horse. When the diaphragm muscles contract, the dome is pulled flatter and this increases the size of the chest cavity. This diaphragm activity is important for deeper breathing. However, if the stomach and intestines are full, the movement of the diaphragm is impeded. This is why a racehorse has no hay and only a small feed in the morning on a race day.

When the horse is relaxed and at rest, the amount of air taken into the lungs is about one-fifth of the amount taken in when the horse is at full exertion. At rest, the horse usually has a respiratory rate of 8 to 16 breaths per minute, the rate being higher for younger stock. After prolonged rapid exertion this rate may be increased to 120 breaths per minute. The lungs never empty totally, but take in and expel more air when the horse is exerted.

Disorders of the Respiratory System

'Broken Wind' or Obstructive Pulmonary Disease (O.P.D.)

Symptoms: Frequent coughing, often spasmodic. The coughs sound long, deep and hollow. There may be respiratory distress. At the end of respiration a line from the stifle may be seen along the belly and there is a double lift on expiration. Breathing is laboured and wheezy.

Causes: It is an allergic reaction; the most common causes are the fungal spores found in hay and straw.

Treatment: The allergy is generally incurable; the symptoms can be treated successfully to alleviate the condition using a drug like 'Intal' for humans given by powered equipment. Recurrence can be prevented in many cases by drugs and good management of the environment. If the coughing is allowed to continue, it may damage the tissues in the lungs (emphysema). Management is based on minimum dust, using paper bedding and haylage.

Chill or Cold (Upper Respiratory Tract Diseases, U.R.T.)

Symptoms: A clear, thick discharge from the nostrils which later may become whitish; swollen throat glands; a gentle or wheezy cough; possibly some difficulty in swallowing and a slight rise in temperature.

Cause: A virus causes the primary infection. The horse standing in an area which is poorly ventilated, such as a stable or lorry, or when first stabled after being out at grass, is more susceptible to the virus.

Treatment: Isolate the horse, keep it warm but allow plenty of fresh air, use paper bedding or arrange for the horse to be out of the stable when the bedding is shaken up or fresh bedding is laid. Hay, if fed, should be soaked and fed on the ground; give soft food and green food. An antiseptic electuary may be given. An ointment designed for human nasal congestion and 'chestiness' can also be rubbed not on the chest and throat but into the outer nostril, twice a day. Inhalations of friar's balsam may also be used. The condition will generally clear up in a week and will respond well to antibiotics, which control secondary infections. Keep the horse on walking exercise until the condition is completely cleared.

Prevention: Avoid stuffy conditions for horses and take particular care when getting them up from grass. Do not let horses drink from public troughs at shows. Avoid allowing the horse to stand in a draught, especially when it is cooling.

Influenza or 'The Cough'

Symptoms: Generally similar to a cold. The horse may first appear shivery or off its food. Raised temperature.

Cause: A virulent virus which has different strains.

Treatment: Call the vet without delay. Antibiotics help. Proceed as for a cold or chill. Generally, avoid hard or fast work for one week for each day of raised temperature.

Prevention: Once the virus is in the yard it is nearly impossible to stop it going from one horse to another. When 'the cough' is known to be in the area, it is, therefore, safest to avoid contact with other horses. There is a vaccine which gives protection against many but not all of the flu strains. Vaccination is compulsory before attending many competitions.

Laryngitis, Tracheitis, Bronchitis

Symptoms: Coughing. A drop in temperature. Discharge from the nostrils. Difficulty in breathing. All three diseases are similar.

Cause: Infection of the upper respiratory tract.

Treatment: Keep the horse warm and give plenty of fresh air. Damp the food before feeding. Electuaries, antibiotics and steaming may help.

Lungworm

Symptoms: A dry cough. Laboratory examination of droppings is necessary to confirm an attack.

Cause: Worm larvae.

Treatment: Call the vet. Fit animals are most resistant to infestation.

Prevention: Keep the pasture clean.

Note: Some horses and most donkeys can carry the disease without showing symptoms.

Nose Bleed (Epistaxis)

Symptoms: Blood appearing at the nostrils. The bleeding may occur from the lungs following fast work. It may come from the guttural pouch, the top of the throat, or from the nose following stomach tubing.

Cause: May be associated with viral infection.

Treatment: Keep the horse quiet; often the bleeding will stop of its own accord. The vet may be able to offer a line of treatment.

Pneumonia

Symptoms: Rise in temperature up to 41°C (107°F). After 12 hours, quickened pulse and respiration; cold extremities. The animal is 'tucked up'. Coughing.
Causes: Inflammation of the lungs caused by a virus, bacteria, a fungus or a parasite.
Treatment: Call the vet. Keep the horse quiet and warm, with plenty of fresh air. Steam inhalation may help. Long convalescence.

Sinusitis

Symptoms: Grey nasal discharge, generally from only one nostril. The area below the eye on the affected side may be swollen and tender.
Causes: Infection, often following a cold or strangles. A diseased tooth.
Treatment: Call the vet who may drain the sinus. Antibiotics and sulphonamides will help.

'Wrong in the Wind' (Whistling, Roaring, 'Makes a Noise')

Symptoms: Abnormal noise as horse inhales at canter or gallop.
Diagnosis: Some cases are clear cut and easily identified; others are a matter of opinion and the experts may agree to differ as to whether a horse is unsound in the wind, or has a temporary infection which affects respiratory function, or is merely unfit and rather 'thick in its wind', which will come right with improved fitness. Viewing with a laryngoscope may show if there is some lack of vocal cord movement.
Cause: Usually, it is the nerve to the left side of the larynx which malfunctions; thus that vocal cord and its neighbouring cartilages are not drawn back during inspiration and the noise is created by air rushing past this obstruction.
Treatment: This problem used to be treated by 'tubing' (tracheotomy), which allows air directly into the windpipe and by-passes the obstruction. Commonly now the horse is 'hobdayed', which is a simple operation to clear the obstruction. The wound after the operation should drain freely and heal well, during which time the horse is kept stabled and quiet. A modification of the Hobday operation includes a prosthesis, which is a man-made spare part.
Prevention: As big horses seem more susceptible to this trouble,

the very greatest care must be taken when getting them fit. The British Ministry of Agriculture lists the disease as hereditary but not everyone would agree. Horses that have been hobdayed are not eligible for show hunter classes.

7 The skin

The skin is many things. It is the body's outer protective (integumentary) layer or covering and the largest organ of the body. It is a very tough and complex packaging system. Horse hide, for example, is used to make leather suits to protect motorcycle speedway riders, who frequently crash at high speed on an abrasive surface.

This envelope to contain the horse is also a protective covering for the underlying tissue, but has several other functions. It stabilises body heat, eliminates waste products in solution with the sweat, and provides camouflage for the horse in the wild. The skin also has specialised nerve endings through which the animal receives the sensations of touch, pressure, cold, heat and pain, as explained in chapter 4.

The thickness of the skin varies with both breed and area of the body. Thus, the skin of the Thoroughbred and similar types such as the Arab is thinner than in the less aristocratic breeds, and the skin across the back, for example, is thicker than that of the face.

The skin is also an indicator of health. Movement of the skin over the ribs shows the presence of subcutaneous fat even in a fit horse. Similarly, a pinched-up fold of skin may give an indication of dehydration if the fold is slow to disappear.

Structure

Skin

Skin (see Fig. 7.1) is the tissue forming the outer covering of the horse's body. It consists of two layers: the inner layer, known as the dermis, and the outer layer or epidermis.

The epidermis is covered with hairs, which form the coat, and it is modified at the extremities of the horse's limbs to form the

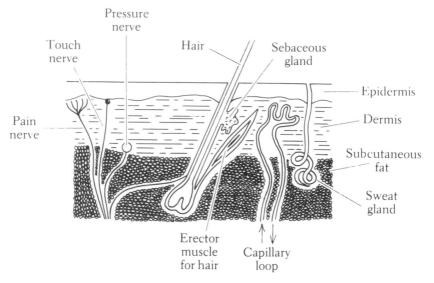

Fig. 7.1 Skin.

hooves. It is a superficial layer which is being shed constantly in the form of scurf because, like most cells within the body, its cells are continually dying and being replaced. The scurf needs to be groomed from the body so that it does not litter the surface and impede some of the skin's other functions.

Horses in the wild do not need grooming. It is not in their nature to dash around working up a sweat. In the wild, the dead cells mix with grease, mud and so on, and the accumulation falls off gradually. In contrast, the ridden horse is being asked for sustained activity and high performance. Regular grooming increases the power of the skin to work at high pressure as well as being an aid to cleanliness and a preventive of disease.

The dermis or inner layer of the skin is deep and sensitive. It contains blood vessels, nerve fibres, glands producing sweat and oil, and hair roots. The hairs and the tubes of the oil and sweat glands pass through the epidermis to the surface, where their openings are known as pores.

The skin is very elastic, varying in thickness according to the amount of protection required. It moves freely over the horse's flesh and should feel loose when handled. The inner layer is seated on a thin layer of subcutaneous fat.

The sweat glands lie deep in the skin, and are little coiled tubes continuing to the surface by a thin duct through which the sweat is discharged. They are in action constantly although this will not be visible when the horse is at rest.

Hair

Almost the whole of the horse's body is covered by hairs. They grow from the hair bulbs, set deep in the skin, and come out at an acute angle with the surface so that the coat lies flat and smooth. Hair is shed, a little at a time, constantly throughout the year, being replaced by new growth. The horse sheds its coat twice yearly, in spring and summer, the thickness of the new growth depending on the season. Each hair is lubricated by oil which exudes from a small gland at the base, and has a tiny muscle which can pull it into an upright position. The hair is part of the mechanism for stabilising body heat. Some hairs, such as the whiskers on the muzzle, are modified so as to act in a sensory capacity, rather like antennae.

The colour of the hair is due only to a single pigment which, by its variation in quality and grouping, gives the colour range we know. This ranges from grey, through chestnut to black, or almost any combination of these colours. There is no pigment in white hairs. A roan-coloured horse has a mixture of white and coloured hairs, whereas a dun has a diluted hair colour owing to the pigment being grouped to one side of the hair. In an albino there is no pigment in either the hair or the skin: the pinkness comes from the blood below.

Coat Colour

When describing horses, it is useful to use agreed terminology about coat colour. In some cases, description is not easy, but attention to the muzzle, mane and stifle area, and a check on the skin colour, may provide the answer. Small patches of white hair on various parts of the body (called 'marks' or 'markings') are ignored when assessing the colour of the coat. Marks must be referred to when giving a full description of the horse.

The following classification is arranged in order of colour density.

Black: This must be total and without any brown hairs.

Brown: A deep and hardy colour, with a mixture of black and brown in the coat, but limbs, mane and tail will all be black.

Bay-brown: A brown horse with a bay muzzle. Although this is the correct term when describing a Thoroughbred, horses of this colour are usually called dark bay in other breeds.

Bay: This has a black mane and tail. Typically, it has black on the limbs. The body colour ranges from bay-brown to light chestnut.

Chestnut: If the mane and tail are not black and the body is in the brown to gold colour range, the colour is described as chestnut. It ranges from the brown horse, with brown mane and tail (liver chestnut), to the light-coloured chestnut with a flaxen mane and tail, which borders on the Palomino.

Roan: Any of the colours mentioned above may be found diluted with a mixture of white hairs. This is called roan. Blacks or browns so diluted become 'blue roans'; bays become 'bay roans' or 'red roans'; and chestnuts become 'chestnut roans' or 'strawberry roans'.

Dun: A black skin with a diluted and evenly distributed coat colour and a black mane and tail. The 'blue dun' has a diluted black coat; the 'yellow dun' has a diluted yellow coat (this is sometimes called 'mouse-coloured'). There may be a dark stripe of hair running the length of the back, which is termed a 'list', 'ray' or 'eel stripe'.

Grey: A varying mosaic of black and white hairs on a black skin producing a range from the colour of lead through to white. A grey horse tends to go white with age. 'Dappled grey' is a mottled effect, and a 'flea-bitten grey' has small clusters of darker hair, often including reddish brown.

Cream: A pale coat on a pink or unpigmented skin, commonly with pink or blue eyes.

Piebald: Black and white in large and distinct patches of each.

Skewbald: Irregular and distinct areas of white and any colour other than black. Where the colours are less clearly defined, it may be described as 'odd-coloured', especially where the patches are of more than two colours.

Markings

Many horses have small patches of hair of a different colour on the body. Most of these marks are clearly defined. Eye colour is sometimes of relevance, and acquired markings (which are patches or adventitious marks), are also used when giving a full description of the horse. The following is a list of the more usual marks.

Blaze: A white marking spread over the forehead and sometimes down the whole width of the face. If the blaze is exaggerated, the term 'white face' is sometimes used.

Flesh marks: Patches of skin with no colouring matter.

List (or ray): The dark lines seen along the backs of some horses.

Snip: A white or pinkish patch on either nostril or lip.

Sock: White hair extending only a short way up the leg.

Star: A patch of white hair on the forehead.

Stocking: White on the leg, extending from the coronet to the hock or knee.

Stripe: A white mark down the front of the face, less wide than the nasal bone.

Wall-eye: An eye of bluish-white appearance, because of a lack of colouring in the iris.

Whorl: A patch of hair lying against the normal line of the coat.

An international standard for description of horses is set out in an *International Guide to the Identification of Thoroughbred Horses* (published by the French Racing Authority), and in a Report published by the Royal College of Veterinary Surgeons (revised edition 1954). Note that certain of the descriptions in the foregoing list are traditional (namely sock and stocking) and may not be used in veterinary practice.

A full description would contain the following details of the animal: name, age, colour, breed or type, sex, height, parentage, natural marks (head, body, limbs), whorls, acquired marks. In referring to limbs, the RCVS Report recommended that the use of the terms 'near' and 'off' should be discontinued and that the terms 'left' and 'right' should be used exclusively, but this recommendation has not found universal acceptance.

Protection

The colour of the skin and the coat gives protection from sunlight. The skin is not necessarily the same colour as the coat, as in the case of a grey. The albino, being devoid of pigment, is more vulnerable to exposure to sunlight and often has a lower resistance to infection. The coat also gives protection from thorns, brambles and the like. The skin is tough in its outer layer, is attached loosely and is free-moving so as to minimise the risk of its tearing. It keeps water out, and yet is able to expel excess moisture as required. This is done through the sweat glands. The subcutaneous fat under the skin acts as a padding to protect the horse's body from minor bumps.

Stabilising Body Heat

The muscular activity of the body generates heat, which is then lost from the body by radiation and the evaporation of sweat. If the weather is hot or there is intense muscular activity (as in galloping), the blood vessels in the skin expand to radiate more heat and to stimulate the sweat glands to greater activity. This is then noticeable as 'lathering up', particularly in Thoroughbred horses.

The skin is cooled by the action of the sweat evaporating off the skin. Sweat evaporates more slowly in very humid conditions and so the cooling effect is then less. To work a horse hard and long in hot and humid conditions creates for the body the serious problems of overheating and dehydration, which cause unacceptable distress to the horse. In extreme circumstances this can result in death.

The horse needs to conserve its body heat in cold weather and therefore the blood vessels in the skin contract. Less heat is then lost by radiation. When the horse is ill, there is often a battle between the invading germs or bacteria and the body's defences, and this leads to sweating.

The horse's coat keeps the body warm, and the oil produced by the glands under the skin's surface greases the hairs and renders the surface waterproof. In cold weather, the hairs stand on end, so increasing the air trapped in the coat, which also grows longer and thicker, thus providing yet more protection. In hot weather the hair lies flat and is much shorter, replacing the long coat that has been shed in spring. Horses of Eastern origin, such as the Arab and the Thoroughbred, have a shorter and finer coat than British native stock.

When the horse is debilitated, the subcutaneous fat is reduced so the horse gets cold and to cope he tends to make the hairs of the coat stand up or 'stare'. Want of condition, neglect or ill health will also cause the coat to look dull and feel harsh. In good health, the coat lies flat, feels quite smooth and has a good gloss.

Where a horse is clipped out, as is the hunter during the winter, the balance must be put right by the provision of rugs, blankets and food. Clothing on horses stops the hair growing so quickly.

Waste Disposal

The function of eliminating waste products in solution with the sweat is particularly important for the horse under stress. Where horses sweat a lot from other reasons, it is evident that they will get rid of more salts from the body than are surplus to requirements. In such circumstances, therefore, greater care must be taken with nutrition to make good the deficit. Extreme cases of such losses of essential salts (electrolytes) may occur in three-day eventing or long-distance riding in hot weather. On completing their activity such horses may need a solution of these salts that are required by the body.

For this aspect of the system to work efficiently, it is important that the pores in the skin be kept open. They have a tendency to become clogged with dirt, and debris such as dead skin cells. The horse in work therefore requires efficient use of the body brush. This is most effective when the horse is still rather warm after exercise, as the pores are then open.

Camouflage

The horse's coat colour is a means of camouflage in its wild state, enabling it to escape the notice of its enemies by allowing it to blend into the background. This can be seen in the wild ponies on both Exmoor and Dartmoor, where the coat colour tends to blend in with the bracken and other background foliage. The striped coat of the zebra, which is a member of the horse family, is another example of camouflage.

Wounds and Healing

Wounds may take several forms: cuts, tears, lacerations and puncture wounds. All of them, but especially the last, can allow tetanus bacteria to enter the body. Unless the horse has been immunised, it may need antitetanus serum if wounded.

The normal process of healing is for the proud flesh to fill the gap up to the surface and then the tissue (epithelium) will grow across forming a union under the scab. When the wound is healed, the scab falls away, often leaving a scar. This is known as healing by second intention. The alternative is to suture (stitch) the skin edges together so that the tissue can reunite without the formation of scar tissue (healing by first intention).

Any wound that is to be sutured should have the stitches put in on the day of the accident, as scar tissue will start to grow after about 12 hours. Dressings on wounds may offer protection for the bruising and may help to keep dirt out; they should not be air-tight because air aids the healing process.

Disorders of the Skin

Skin diseases and disorders are a common occurrence in horses.

Cracked Heels

Symptoms: An inflamed area with hard skin and red raw cracks, causing pain and lameness.
Cause: Allowing the heels to get wet, scratched, or muddy, and not attending to them properly at the end of the day. This complaint especially affects clipped-out horses.
Treatment: Acriflavine cream rubbed in gently each day. The scabs must be removed.
Prevention: When washing the feet, put the thumb in the deep groove in the heel to ensure that it is cleansed. Grease heels with udder cream or similar protection before going hunting in wet conditions. Dry heels gently when wet and do not brush out mud with a stiff brush.

Girth Gall

Symptoms: A sore area under the girth.

Causes: Sensitive skin. Girth causing rubbing, especially a dirty girth.
Treatment: As for a cut. By placing a cotton wool pad over the area, the horse can normally be ridden again after a few days. Exercise in hand in the meantime.
Prevention: Surgical spirit can be used to harden the area once healed.

Mange

Symptoms: Intense skin irritation. Small lumps on the withers, back, neck and shoulders.
Cause: Parasites (mange mites).
Treatment: The vet must be informed as some forms are notifiable to the authorities. An affected horse should be isolated. (*Note*: this is a very rare disease in Britain.)

Mud Fever

Symptoms: Inflamed skin, often with some degree of swelling. Frequently found in the heels of horses in wet weather. The legs and sometimes the belly are affected. Mud fever often goes with cracked heels.
Cause: Wetting or chapping by mud or water predispose to invasion by a microbe found in some soils.
Treatment: In severe cases, cut food and give light laxative diet. Poultice the legs to draw out infection. Antibiotic cream or waterproofing the skin with liquid paraffin helps.
Prevention: Do not brush the belly and legs of a tired, muddy horse. Either apply stable bandages and clean next day, or wash clean with a sponge and cold water. Dry gently but thoroughly, especially the heels.

Lice

Symptoms: Skin irritation in the mane, neck and side of the chest, especially in spring and early summer.
Cause: Lice.
Treatment: Dust with louse powder. Check general condition and improve diet if necessary. Clean coat thoroughly.

Rain Scald

Symptoms: Rough, scabby areas on the back and quarters.
Cause: An organism which gains entry to the upper layers of the

skin when the coat has been soaked in water for long periods.
Treatment: Dust with sulphanilamide powder. Use antiseptic soap.
Waterproof the skin with liquid paraffin. Provide a shelter.

Ringworm

Symptoms: A circular area of raised hair with exuding fluid, and
later a bald patch.
Cause: Fungal infection, which can be carried by the grooming kit,
tack, clothing, etc.
Treatment: Isolate the horse. The vet will prescribe a suitable drug.

Sit-fast

Symptoms: A dying lump of flesh under the saddle patch.
Cause: Severe pressure from a saddle.
Treatment: Poultice until the dead area comes away, then bathe
clean daily and apply antibiotic ointment. The vet's help may be
needed.
Prevention: Use a properly fitted saddle.

Sore Back

Symptoms: Tender back. Sometimes there are sores.
Cause: Loose girth. Poorly stuffed or dirty saddle. Bad rider.
Treatment: As for girth gall (above).

Sweet Itch

Symptoms: The mane, back, quarters and tail look inflamed and
rubbed. Sores.
Cause: Allergic dermatitis. Midge bites make certain horses
allergic to some proteins in their diet. It is found particularly in
ponies, generally in spring and summer, on animals at grass.
Treatment: Stable at dawn and dusk. Keep the affected area clean.
Apply the parasiticide benzyl benzoate every other day. At first
occurrence, get the vet to check for mange.

Urticaria (Nettle Rash)

Symptoms: Raised areas or swellings on the skin, caused by
accumulation of fluid under the skin.
Cause: There are several possible causes. It may be an allergic
reaction to plant bites or stings, but it sometimes indicates an
excess of protein in the diet.
Treatment: A day's laxative diet, such as a bran mash with a little

Glauber's or Epsom salts, then use calomine lotion or 2 tablespoons of bicarbonate of soda in half a litre of water, applied externally. Antihistamines may be helpful.

Warbles

Symptoms: Firm nodules or swellings on the back. After a fortnight or so the swellings are bigger and may have a hole in the top.

Cause: Warble-fly larvae. The fly lays its eggs on the coat, and the larvae eventually make their way to the skin, usually in spring and early summer.

Treatment: It can be dangerous to try and squeeze out the larvae. If they are squashed under the skin, for example by the saddle, a massive allergic reaction can occur. The condition is therefore best left alone. Poultices should not be used except when a grub has been squashed and an acute reaction has occurred.

8 Digestive system

The function of the digestive system is to process and extract most of the nutrients from whatever foodstuffs are put into it. The system consists of the teeth, the accessory organs, and the alimentary canal, which runs the length of the horse's body.

The digestive tract starts at the mouth and ends at the anus. Food and other materials passing along it are not part of the horse's body, but are merely travelling through it. Nutrients are extracted and waste matter is voided. The digestive tract is both lengthy and complex, and if the horse is fed contrary to its natural requirements, problems will arise. The tract tends to be prone to invasion by parasites and relies on the horse and its keeper to be selective over what is put into it.

Teeth

The teeth provide a cutting and grinding system which is both simple and yet highly specialised. The horse's teeth (Fig. 8.1) are divided into the cutting teeth at the front (the incisors), and the grinding teeth at the back (the molars). Horses, like human beings, have two sets of teeth: in infancy they have milk or deciduous teeth and later these are replaced by a second or permanent set of teeth.

The molars are unique in their adaptation for grinding coarse hard herbage. They have big surfaces or 'tables', covered with complicated patterns of enamel. As the horse chews by moving its jaws from side to side, the molars grind the food; they also wear against each other. Growth balances wear. However, because the lower jaw is narrower than the upper, the wear on the tables is sometimes uneven. Small areas at the sides of the tables do not get worn down and they stick up as sharp edges. This may cause sores

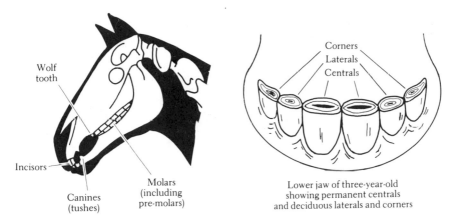

Fig. 8.1 Teeth.

on the cheeks or the tongue. Potential areas of trouble are the inside edges on the lower jaw and the outside edges on the upper jaw. Rasping (see chapter 2) will deal with this problem.

With experience, a horse's age can be judged by examining its teeth. The first essential is to be able to distinguish the milk teeth from the permanent teeth. At the deciduous stage, the incisors are whiter, smaller and seem to be pointed at the gum. The incisors are called centrals (the pair top and bottom in the centre), laterals (the teeth next to the centrals), and corners. (See Fig. 8.1.)

In judging the age of foals it is easier to look at other factors such as the growth of the tail and the stage of development. At a year old, there are six neat white incisors both at top and bottom, but the corners will hardly meet. At two years, the incisors are similar to those of the six-year-old except that they are milk teeth. At three, the two centrals have been replaced by permanent teeth. When the horse is four years old, the laterals will have been similarly replaced. By five years of age, the corners have followed suit; however, the corner teeth will only be touching their opposite number at the front edge. Tushes may have emerged, but they do not always appear in mares and geldings. By six years of age, the corners still are not in complete wear across their tables when compared with the laterals and centrals. Moreover, when looking at the tables, the centre of each tooth will be seen to have a dark hollow area (known as the mark). At seven, the marks of the centrals and laterals are smaller than on the corners, and the top corner teeth overlap the bottom ones so that there may be a little

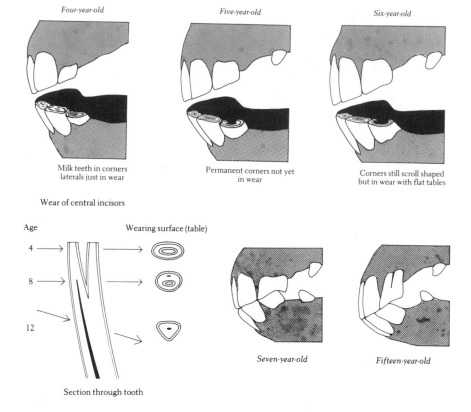

Four-year-old

Five-year-old

Six-year-old

Milk teeth in corners
laterals just in wear

Permanent corners not yet
in wear

Corners still scroll shaped
but in wear with flat tables

Wear of central incisors

Age

Wearing surface (table)

4

8

12

Seven-year-old

Fifteen-year-old

Section through tooth

Fig. 8.2 Ageing.

projection on the trailing edge. This is known as the seven-year hook. By eight the marks are the same on all teeth and the 'hook' has disappeared. (See Fig. 8.2.)

Galvayne's groove appears on the upper corner teeth at nine years of age. This is a dark area just below the gum and extends downwards over the next ten years, so that at 15 it will be half-way down the tooth. Another change over the next ten years is that the angle of the teeth becomes less upright. Gradually, the incisors point further forwards and appear to get longer. The third change over the next ten years is in shape. The table goes from long-oval to triangular.

Accessory Organs

Liver

The liver is the largest gland in the horse's body and lies up against the diaphragm. It is said to have over a hundred different functions. It supplies a fluid called bile to aid digestion, and also stores energy in the form of glycogen. The liver also converts amino acids into proteins. It regulates the blood and controls the nutrients carried in it. Because it monitors all that the blood carries, its functions may be impaired as a result of damage caused by infections, toxins and other poisons. In some cases this damage is cumulative, for example through eating ragwort. Inflammation of the liver is called hepatitis, and jaundice is a symptom of various forms of liver malfunction as a result of damage or disease.

Pancreas

This large gland produces digestive juices. It also produces the hormone insulin to control blood sugar levels.

Digestive Tract

The digestive tract of the horse is shown diagrammatically in Fig. 8.3. The horse's lips are specially mobile and adapted for grasping. Its incisor teeth are used for biting off herbage, in contrast to the teeth of cows where the tongue is used in conjunction with the teeth meeting a hard pad. In the mouth, the food is moved by the tongue which transfers it from the front towards the back and sides so that the molar teeth can grind it. During this process the salivary glands (the parotid, and mandibular and sublingual glands, arranged in pairs), produce saliva, which wets and warms the food. Finally, the tongue forms a portion of food into a bolus, which is rolled off the back of the tongue. This triggers the pharynx to operate the epiglottis so that the bolus passes through the pharynx, entering on the lower channel and leaving on the upper to go along the oesophagus or gullet. This is a tube 4 to 5 ft (up to 1.5 m) long, which goes from the back of the throat down the neck. Food being swallowed can be seen on the left side of the neck passing down just behind the windpipe or trachea. The oesophagus goes through the chest, passing between the lungs, and on through the diaphragm into the abdominal cavity where it enters the stomach. The abdominal cavity is lined with a smooth membrane (the peritoneum), which also covers the intestines and helps them slide easily against each other like a tub full of writhing snakes. Food is moved along the gut by waves of muscular contraction called peristalsis. Movement down the oesophagus is aided by the

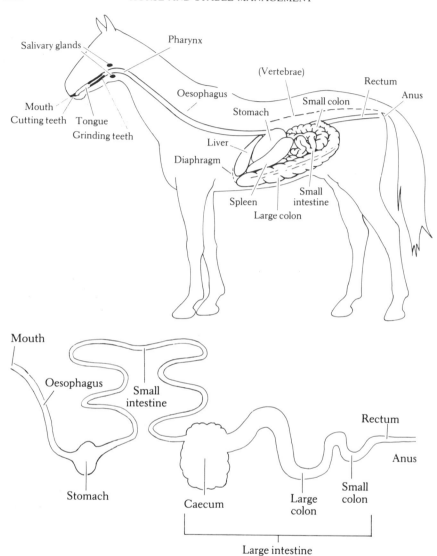

Salivary glands
Pharynx
(Vertebrae)
Rectum
Oesophagus
Anus
Small colon
Mouth
Stomach
Cutting teeth   Tongue
Grinding teeth
Liver
Diaphragm
Small
intestine
Spleen
Large colon

Mouth
Oesophagus
Small
intestine
Rectum
Anus
Stomach
Caecum
Large
colon
Small
colon
Large intestine

Fig. 8.3 Digestive system and alimentary canal.

lubrication provided by the saliva.

The empty stomach is only about the size of a rugby football but fills to contain 9–18 litres (2–4 gal). The ring muscle controlling the inlet into the stomach is called the cardiac sphincter; the muscle at the outlet is the pyloric sphincter.

The stomach is followed by the small intestine, which has three parts. The first metre (3 ft) is called the duodenum, into which flow the ducts from the liver and pancreas. The main part of the small intestine is the jejunum. This is nearly 20 m long and its final part (about 2 m long) is called the ileum. These three parts hold about 50 litres (12 gal). The diameter of this long tube is greater than that of a garden hose but smaller than that of a fireman's canvas hose. The small intestine is held up in loose coils by the mesentery that is formed by the peritoneum being folded or looped around the gut and attached to the roof of the cavity. It carries blood vessels and nerves for the gut.

The small intestine opens into the large intestine which, although only half as long as the former, has nearly three times the volume. The first section is a blind gut called the caecum. It holds about 35 litres (8 gal). The next section is the large or great colon, 3 to 4 m long and holding up to 80 litres (around 16 to 20 gal). The small colon is about the same length, but it is only about the same diameter as the small intestine; it only holds up to 16 litres (about 3 gal). The final third of a metre is called the rectum, which ends in the anus where there is another sphincter muscle.

Unfortunately, the horse's digestive tract has design faults. The horse eats bulky cellulose foods such as grass, and these are not broken down until they reach the large intestine. Thus, the food passes through the small intestine in a much coarser state than it does in most animals. In man, for example, there is a comparatively large stomach, which is able to cope with food breakdown; the stomach of a cow is large and able to break down cellulose. Because of these imperfections in equine design, where a section of small intestine is suffering from more than usual worm damage, colic may occur as a result of the pain of restricted flow or dead tissue. Similarly, because semi-solids do not flow as well as liquids and because, in order to get the gut into the cavity, it is turned through sharp bends, blockages may occur, resulting in colic.

Digestion

The complex foods that the horse eats consist of water, vitamins, minerals, carbohydrates, lipids and proteins. These last three form the bulk of the digestible food. The carbohydrates have to be broken down into simple sugars, which are the source of usable energy. Lipids, which are fats and oils, are broken down to fatty acids and glycerol, which may be used or stored. Proteins are broken down into amino acids which are the building blocks from which much of the body is made.

Little digestion occurs in the mouth because horse saliva, although copious in quantity, is low in enzymes, the chemicals that create changes of state in foods. The food is mixed with gastric juice in the stomach. This is secreted by glands in the wall, and acidifies the food by means of hydrochloric acid. It also contains three enzymes: pepsin, which starts breaking down proteins; renin, which coagulates milk in foals; and lipase, which starts work on lipids.

The horse takes 24 hours to empty a full stomach almost entirely, but it prefers to keep it about half full. The food leaves with the most liquid portions first: water, carbohydrates, proteins and fats, in that order. Five hours after eating a full meal, a horse will usually feel hungry again. Provided the duodenum is empty and the food in the stomach is sufficiently acid, the food is passed on from the stomach to the small intestine.

Secretions from the liver and the pancreas flow into the duodenum. The liver produces bile, which flows down the bile duct and emulsifies lipids, thus aiding their digestion and absorption. Bile turns the acid contents of the stomach into an alkaline mixture to go down the intestine. The horse has no gall bladder: it is meant to be a continuous feeder. Feeding little and often is therefore better for good digestion and the efficient use of food.

The pancreatic juice is alkaline and contains sodium bicarbonate to counter the acidity of the stomach. It also contains enzymes, including trypsin (which breaks down proteins into peptides and then into amino acids) and amylase (which breaks down starch into maltose which, in turn is broken down into glucose by the enzyme maltase).

Digestion continues along the main length of the small intestine (the jejunum). At this stage the food of most animals is a creamy smooth mixture. In the horse, however, this mixture still contains the coarse fibre which is an essential part of the horse's diet. Peristalsis, the process of muscular contraction that moves the food along, also mixes it with the digestive juices and forces it against the intestinal walls where it can be absorbed. The intestinal wall has little protuberances (villi) all along it, which are like the pile on a carpet and thus give a greater surface area. Here the amino acids, glucose, minerals and vitamins pass into the blood stream and some of the fatty acids and glycerol pass into the lymphatic system. Between the villi are little crypts which produce further juices to aid digestion. There is no clear division from the jejunum into the ileum, but there is a valve to control flow into the caecum.

The caecum acts as a holding chamber to keep the great colon topped up, although further digestion can occur there. The breakdown of food in the great colon may take several days. This is why it is so bulky, and why the grass-fed horse has a big belly. To enable it to carry this considerable weight, the horse has a strong back. Because the great colon is so large, at one point it turns sharply back on itself at the pelvic flexure; this sharp turn is sometimes a site of blockage. Bacteria live in the caecum and the colon, and break down cellulose to release its volatile fatty acids. These bacteria are of many types and have very short lives. They are fairly specific to different foods so it is important to change food sources gradually so that the population of bacteria can also change to adapt to the new diet. Besides breaking down foods, the bacteria can also build them up into essential vitamins.

Food passes from the great colon into the small colon, where nutrients and water are still being extracted. It reaches the rectum where further water is removed. The waste material is formed into balls of dung or faeces, which are evacuated at intervals through the anus. On average, the food takes three to four days to pass through the horse. Examination of faeces after feeding whole grain will show that some foods pass through more quickly than this, and indeed some grains remain intact.

Parasites of the Digestive Tract

Flatworms

Flatworms are rarely important. The *liver fluke* (which is a form of flatworm) is found in wet conditions. To complete its life cycle it has an intermediate host, the mud snail. Control of an area includes draining land and killing the snails. The liver fluke normally attacks cattle and sheep, but where a horse is invaded it will cause unthriftiness, and will stunt growth and lead to anaemia.

The other flatworm to note is the *tapeworm*. Tapeworms can grow to 80 cm (2½ ft) long, but are usually 5 to 8 cm (2 to 3 in) long. They live in the large and small intestine. Their intermediate host is a mite. They rarely produce symptoms but may cause unthriftiness.

Roundworms

The three important roundworms that affect the horse are the seat- or whipworm, the large round whiteworm and the red bloodworm.

The *seatworm* (*Oxyuris*) female lays eggs at the horse's anus. Larvae develop in the eggs, drop to the ground and get eaten with food. The small (up to 15 cm (6 in) long) adult worms live in the caecum and colon. The worm causes the horse to have an itchy anus and may lead to tail rubbing. The yellow eggs may be seen round the anus. Routine worming controls this pest.

Whiteworms (*Ascarids*): these large round worms are up to 1 ft long (15–30 cm) and as thick as a pencil. They live in the small intestine. The female lays eggs (at a laying rate of 8000 per hour per worm), which are passed out in the droppings. In favourable conditions on pasture, eggs hatch to form larvae which remain inside the eggshell for protection. The eggs are infective for a period of from 30 days to up to 3 years. When eaten the larvae go through the gut wall and migrate via the liver and heart to the lungs where they get coughed up and reswallowed. They then become adults.

Whiteworms cause some loss of condition for horses under three years old. Principally they are a foal problem. Adult horses that have whiteworms must be treated if foals have access to their pasture.

Redworms (*Strongylus*) are tiny worms, as thin as cotton and as long as a fingernail. They live in the intestines, where the female lays eggs that pass out with the droppings. Redworms can be extracted from the droppings after 'worming', and their eggs can be identified with a microscope. The eggs hatch when conditions are suitably warm and moist. The third-stage infective larvae crawl up blades of grass and get eaten. They then burrow into the gut linings, and although some complete their development there in three or four months, others travel further. Some invade the liver after an eight-month journey. The larvae of one species (*Strongylus vulgaris*, the large redworm) get into the arteries supplying the intestines and travel against the blood flow until reaching the anterior (cranial) mesenteric artery (see Fig. 8.4). This is the main blood vessel supplying the intestines. They develop there for several months before returning down the blood vessel and back into the intestine. These expeditions lead to damaged artery walls and blood clots where the worms settle.

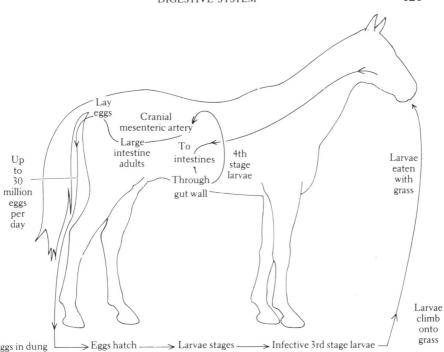

Fig. 8.4 Life cycle of large redworm (*Strongylus vulgaris*): 200 days.

These clots (thrombi) may break off and block blood vessels, thus stopping blood supply to the part of the intestine served by that vessel. The symptoms of this worm damage may include loss of condition, anaemia, distended stomach, staring coat, diarrhoea and colic. Control includes removing droppings from the paddock, grazing with cattle and sheep, and rotating grazing. Drugs used against worms are known as anthelminthics and some can now kill the wandering larvae as well as the adult worms. Thus, when dosing the horse, there can be a more complete cleansing of the system from these parasites. This possibility has changed the pattern of worm control.

Bots

Bots (*Gastrophilus*) are a non-worm parasite, the larval stage of the gadfly. The adult female lays its eggs on the coat of the forelegs from June to September. The eggs hatch to larvae, which the horse licks off. These burrow into the tongue or cheek, where they remain for 2 to 3 weeks; they then migrate to the stomach and attach themselves to the stomach wall. In spring they let go and pass out with the droppings, maturing into flies. They cause unthriftiness. The eggs should be wiped off with a paraffin rag or scraped off with a knife. Anti-fly aromatic protection (such as a daily dab of oil of citronella) on the forelegs is beneficial. A drug can be used in October and November. One drug available is sprayed on inert little pellets, which are mixed with the food. These are passed out in the droppings, having done their job in conveying the drug into the horse. Note that such droppings will kill poultry or indeed any bird that may eat them.

Disorders of the Digestive System

Choking

Symptoms: Dribbling, attempts to swallow, head repeatedly down to chest with tensed neck.
Cause: Potato, apple, carrot or lump of dry feed.
Treatment: Keep the horse calm. As long as it can inhale air it will not die. *Do not drench or give water*. Proceed as described in First-aid Procedure (chapter 2).
Prevention: Do not give concentrates to a very hungry horse. Give a small drink and a little hay first. Horses cannot vomit, though they can pass liquids back through their nose. Sugar beet nuts should be soaked thoroughly before feeding.

Colic (Tummy Ache)

There are several types of colic, as follows:

(a) *Impaction or stoppage*, caused by food blocking the alimentary canal. It can even occur following dosing against worms, as the expelled worm bodies block the gut.
(b) *Flatulent or tympanic colic*, caused by gas being created faster than it can be absorbed or passed out. It is natural for gas to be created during digestion, but if it is trapped it can distend the gut wall thus causing pain.
(c) *Spasmodic colic* is a condition of an irritated gut wall getting overactive. It is not good for any part of the horse to get 'up-tight', and the vet can use drugs to relax and reduce spasms.
(d) *Artery blockage* by thrombosis caused by bloodworms. Without its blood supply the bit of gut involved dies unless

an alternative route develops in time.

(e)   *Twisted gut* can occur where the mesentery has become torn.

Symptoms: Any or all of the following. The horse is in pain, kicks at its belly, and paws the ground with its front feet. It lies down and may stretch out and groan. Gets up and down or rolls. Stamps feet, breaks out in a cold patchy sweat. Attempts to pass water. Passes small quantities of droppings. Breathing hurried and blowing. Belly inflated. Bowel movements slow down or stop. The temperature may rise.

Causes: Faulty teeth causing poorly chewed food. Too much or unsuitable or badly prepared food. Bolting food. Irregularity of meal times or too long between meals. Poor quality food. Accidental access to food, or to food in the wrong form, e.g. short grass cuttings. A sudden change of diet. Excess of cold water to drink when horse is hot. Over- or under-work. Working immediately after feeding. Watering after feeding. Crib-biting or windsucking. Kidney- or bladder-stones. Grass-sickness disease. Taking sand up when drinking from a stream. Parasites – probably the most common cause.

Treatment: The horse should be brought into the stable if outside. It should be encouraged to stale. The bedding should be topped up and banked round the walls, and any obstructions should be removed. Call the vet. Temperature, pulse and respiration should be taken, and the horse should be kept warm with light clothing. Do not feed but offer water. An occasional walk in hand round the yard may help. The vet may use drugs to relieve pain, relax the horse and ease spasms. He may also administer saline solution and lubricant. Immediate surgery is required in the case of a twisted gut.

Constipation

Symptoms: Small, hard or no droppings.
Cause: Poor diet.
Treatment: Soft laxative diet of green food and bran mashes. Purgatives should not be used except on the vet's advice.

Diarrhoea

Symptoms: Obvious.
Causes: Nerves. Sudden change of diet or too much green food.

Too much rich food. Faulty teeth. Inflammation of the gut from irritant food or worms. Infection.
Treatment: Remove the cause where possible. Diet of hay and a little dry bran. Check temperature: if it is high, an infection is the cause.

Parrot Mouth

Symptoms: Horse in poor condition.
Cause: Malformation with short lower jaw (mandible).
Treatment: In severe cases, feed with complete cubes.

Poisoning

Symptoms: Diarrhoea, colic, convulsions, coma, dilation or constriction of pupils, distressed breathing, muscular incoordination, sensitisation, blood in urine, etc.
Cause: Ingestion of a substance that harms the body internally by interfering with function.
Treatment: Call the vet at once.

Sharp Teeth

Symptoms: Quidding, i.e. chewing food without swallowing. Cautious eating, sometimes with head on one side. Lacerations of tongue and cheek.
Cause: Uneven tooth wear leaving sharp edges.
Treatment: Rasping, as described under treatments (chapter 2). Sloppy food for a few days if there are lacerations.

Sore Throat

Symptoms: Poor appetite, cough, running nose.
Cause: Bacterial activity, often as a secondary infection.
Treatment: The horse should be isolated, kept warm and allowed ample fresh air. Inhalations may help. Dusty bedding and food should be avoided. Sloppy food fed from a bowl on the ground. The vet may prescribe antibiotics.

Wolf Teeth

Symptoms: Horse tosses its head when it has a bit in its mouth.
Cause: Natural formation – not always present.
Treatment: Ask the vet about removal.

9 Reproductive, urinary and mammary systems

The reproductive, urinary and mammary systems are interrelated. The reproductive system's task is to ensure the continuance of the species. Like most animals, the female horse is subject both to cycles and to seasons of the year in her sexual behaviour. For reasons of safety, performance and convenience, the male horse is commonly neutered or gelded. Unfortunately, in spite of man's best endeavours and the advances of science, the number of foals reared compares unfavourably with the number of mares going to stallions each year, especially in the Thoroughbred.

The mammary system provides nourishment for the young, and although the udder is more discreetly placed than in cattle, the mare will provide large quantities of milk. Indeed, in some parts of the world she provides milk for human consumption.

The urinary system filters impurities from the blood, temporarily stores them, and then disposes of them.

The Male Reproductive System

The reproductive organs of the stallion (see Fig. 9.1) are designed to create a sperm and place it in such a position within the mare that it can unite with an egg. Sperm is created in the testes, which function best at a slightly cooler temperature than the rest of the body. They are therefore slung in a thin skin 'purse' (the scrotum), which is located between the hind legs for protection. The testes start life in the abdominal cavity but usually have descended down the inguinal canal into the scrotum by birth. Although the testes may reascend, they are dropped permanently by 12 months. Where one or both testes are retained in the canal or abdomen, the horse is called a 'rig' or cryptorchid. A 'rig' may be purchased in mistake for a gelding. Although the retained testis will not

Male

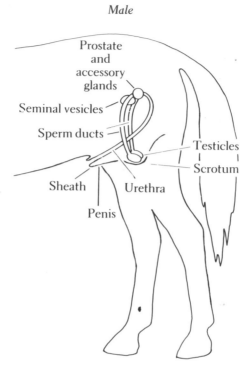

Fig. 9.1 Male reproductive system.

produce fertile sperm, it will create the male sex hormone testosterone, which is responsible for male behaviour. Accordingly, a rig must be treated like a stallion. Rigs may be used as 'teasers' to ascertain whether a mare is in season.

In the mature horse, the testes are round, long, and of a size that would fill a cupped hand. Each testis weighs about 300 g (10 oz.). The sperm created by the testes are stored in small coiled tubes, joining into one, the epididymis. This is attached to the upper edge of each testis. The epididymis has a tube (the vas deferens) which leads up into the body. The two tubes, one from each testis, run side by side as the route for sperm during the sexual act. In the abdominal cavity they lead past the two seminal vesicles which lie either side of the bladder and which, with neighbouring accessory glands, produce seminal fluid. The sperm and seminal fluid together form semen. A stallion will release 40–120 ml (2–5 fl. oz.) of semen at a time, and this will contain about 4000 million sperm. A common duct (the urethra), which carries both semen and urine, runs from the bladder, down to and through the penis, which, at rest, is enclosed in the sheath (prepuce).

When the stallion is sexually aroused, blood flows into the erectile tissue of the penis, which becomes thicker and longer – up to about 20 in (50 cm). When the penis is erect, the stallion can, with practice, insert it into the vagina of a receptive mare. This is called intromission. The head of the penis (the glans) is further enlarged inside the mare, thus securing a tight fit. After some thrusting, semen is ejaculated into the mare and this is accompanied by flagging of the stallion's tail. The penis then starts to reduce in size and the stallion withdraws from the mare's vagina. From mounting to dismounting, the whole operation commonly takes less than a minute. The penis quickly shrinks back into the sheath, which consists of double folds of skin lubricated by smegma. The sheaths of both stallions and geldings need to be kept clean inside and out.

Castration is the operation of removing the testes, and the operation is usually carried out in spring or autumn. A castrated stallion is called a gelding, and it sounds, looks and behaves like a mare out of season.

In general, the male genitalia function well. The vet can take a sample of semen and check it for quality. Occasionally, germs may affect the stallion's penis. They may be picked up from a mare during service. An example is a venereal disease caused by a herpes virus, which causes spots or 'coital exanthema'. This takes the form of blisters on the penis, and necessitates two or three weeks of rest for the stallion. A stallion may also be damaged in use by being kicked by a mare, so causing a swelling filled with blood (a haematoma) on the penis, or one filled with mixed fluids (oedema) of the scrotum.

The Female Reproductive System

The mare's reproductive system (Fig. 9.2) is designed to produce an egg (ovum), which will unite with a sperm to form an embryo. The system also provides nutrition for the embryo and a good environment in which it can develop.

Ova are produced by the ovaries. At birth, these contain all the egg cells needed, and eggs are released at intervals during the

Female

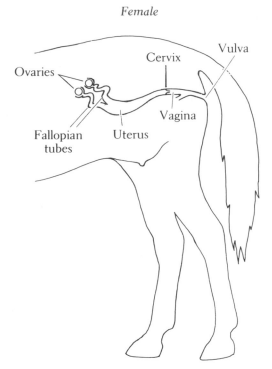

Fig. 9.2 Female reproductive system.

mare's fertile life. The two ovaries are attached high in the abdominal cavity, just behind the kidneys. An expert can feel them through the wall of the rectum. They are bean-shaped and each is about the size of a chicken's egg. An ovum is passed down from an ovary through one of the oviducts (also known as the fallopian tubes), where it may be fertilised. Each fallopian tube runs into a horn of the uterus or womb. This is Y-shaped, with two horns and a main body; it is suspended in the abdominal cavity. At its rear end there is the sphincter muscle (cervix), which closes it for most of the time. A short passage (the vagina) leads from the cervix to the outside. The vagina ends in the vulva, which has two lips or labia on the outside and a small penis-like organ (the clitoris) on the inside.

When the mare is in season, she can flash open the lips of the vulva in an action called 'winking'. This is quite often done after she micturates (passes urine) in the distinctive and slightly squatting position of the in-season mare. Instead of being tightly closed, when the mare is in season the cervix relaxes and opens slightly. When she is covered by the stallion at mating, the mare's vagina takes the stallion's penis. Some of the semen will be lost in the vagina, although most should be ejaculated through the cervix, swim across the uterus and up the fallopian tubes in search of an egg. On finding it, the sperm swim round it trying to gain entry. Only one sperm will succeed and unite with the egg.

Generally, the female genitals function well. However, germs may invade the genital tract and cause problems. They may gain entry during mating (coitus) or because the mare has a poorly shaped vulva which allows air to enter and take in germs. This defect is corrected by stitching using 'Caslick's operation'.

Two diseases which cause – or have caused – great problems are the equine herpes virus and contagious equine metritis. The former results in abortion (rhinopneumonitis) and also causes 'snotty nose' in yearlings. The latter (C.E.M.) is a highly contagious venereal disease causing infertility.

Oestrous Cycle

A mare will reach puberty between 12 and 24 months after birth and is then ready to 'come into season'. Normally a mare comes into season (oestrus) at regular intervals through the summer months, the breeding season. Improved food, lengthening days and warmer sunshine during spring trigger off the process.

The mare's pituitary gland produces F.S.H. (follicle stimulating hormone), which activates her ovaries. One of the ovaries forms a follicle which appears like a hard cyst on the surface. The ovaries also give out the hormone oestrogen (in Greek this means 'to produce mad desire'!). It is this hormone which brings the mare into season (oestrus). The pituitary gland then starts to produce L.H. (luteinising hormone), which stimulates the egg to reach maturity, and then its release (ovulation) into the fallopian tube. This happens when the mare has been in season for about 4 days. The crater on the ovary where the egg was sited now fills up like a soft cyst as the yellow body (corpus luteum). This produces the hormone progesterone. Oestrus then ends and the mare 'goes off', having been in season for about five to six days. The oestrous cycle is summarised in Fig. 9.3.

For the next two weeks the mare has no interest in the stallion.

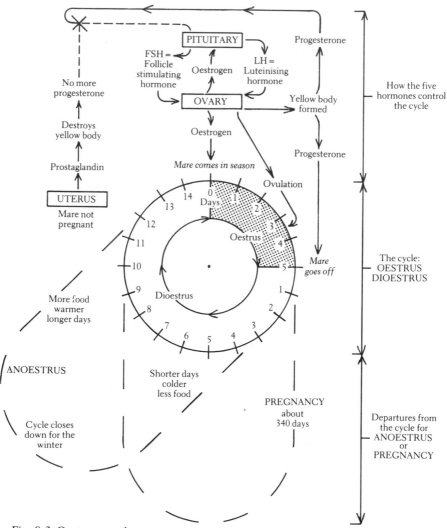

Fig. 9.3 Oestrous cycle.

However, the uterus prepares to receive the fertilised egg. If the egg has not been fertilised to form an embryo, a message from the uterus (the hormone prostaglandin) instructs the corpus luteum to cease its activity and the cycle begins again.

Dioestrus is the period between one oestrus and the next; anoestrus is the period over the winter when the cycle stops. The normal cycle lasts about three weeks and in general is consistent for each mare. By observing a mare's normal cycle it will be easier to predict the time of ovulation.

Pregnancy

The mare is pregnant when the ovum and sperm have united to form an embryo. The embryo grows rapidly by division of cells. Initially, this takes place within the egg capsule which is descending the fallopian tube. After two weeks, the embryo is a rapidly growing mass and is lying in one of the horns of the uterus. After three weeks it has taken on body form, utilising the yolk from the egg. By the sixth week the embryo is well developed and is called a foetus, which is nourished by uterine milk. (See Fig. 9.4.) The foetus floats in amniotic fluid, which is parcelled within a membrane (the amnion). This parcel lies within the allantoic fluid ('the waters') and a second membrane (the placenta). The placenta bonds with the lining of the uterus and is a means of nutrient, oxygen and waste product exchange. It is connected to the foetus by the umbilical cord. Except for the lungs all the systems are working within the foetus. Liquid excreta is passed into the allantoic fluid where some of it solidifies to form a flattish brown piece of matter, the hippomane, which is found among the afterbirth. Physical movement is noticeable in the later stages of pregnancy when the foetus is growing rapidly.

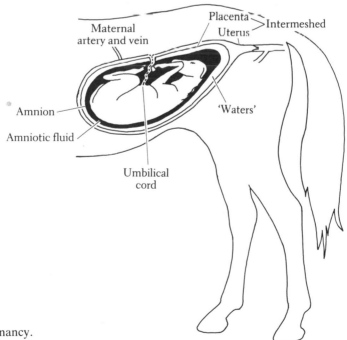

Fig. 9.4 Pregnancy.

Inheritance

There is coded information on the exact make and shape of the parents in an ovum or sperm. This information is contained in chromosomes. Each chromosome is like a string of beads, each bead being called a gene and having a special function to perform, e.g. coat colour. When ovum and sperm fuse to form the embryo, they each have half the total number of chromosomes, so in the embryo the number is up to full strength (64). This total is made up of 32 different pairs, one from each parent. If the genes are the same (homozygous), then there is no conflict. If they are different (heterozygous), then one will prove dominant and the other recessive. Homozygous genes always breed true. This is not always so with heterozygous genes. An example of this is the great racehorse, The Tetrarch. He had only one grey great-grandparent (out of eight). This grey produced a grey (one out of four grandparents), which in turn produced another grey (one out of two parents); this in turn produced a grey, The Tetrarch. This shows dominance of grey coat colour.

In genetics, the gene that results in horses with pricked ears dominates that giving lop ears; similarly the gene for a dished face dominates that for a Roman nose. The genes also determine the sex of the foal, and the sperm always carries the deciding factor.

Recessive genes may show up if *inbreeding* (breeding close relations) is practised. Inbreeding can strengthen the genetic make-up provided there is no history of undesirable characteristics. Close inbreeding includes sire to daughter, dam to son, and brother to sister. Animals born with undesirable characteristics should not be used as breeding stock. *Line-breeding* includes grandfather to granddaughter, grandmother to grandson, and cousin to cousin. Line-breeding has less risks but takes longer to establish purity. *Outbreeding* is where there is no relation within the previous five generations.

Mating outside the breed is known as crossing. A true hybrid results from mating with another species, e.g. a donkey with a pony. Usually, such hybrids are vigorous but infertile, as is the case with mules. Crossing between different breeds can produce some hybrid vigour. Such vigour was found when horses from the south and east Mediterranean were crossed with English native improved stock to form the Thoroughbred (see Figs 9.5 and 9.6).

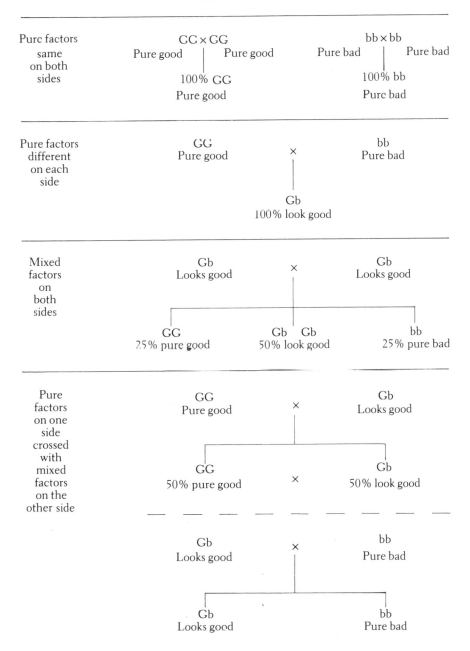

Fig. 9.5 Inheritance of a single trait or character.

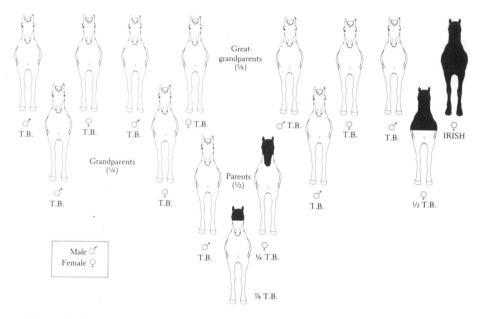

Fig. 9.6 Inheritance.

Mammary System

The mare's udder develops to suckle the foal. The udder consists of the mammary glands in two separate compartments, each leading to a teat. The udder and two teats are located between the hind legs for protection. A big mare can produce up to 23 litres (5 gal) of milk per day.

The mammary glands are well supplied with blood and lymph vessels. The tissues concerned with production are grouped around little sacks called alveoli (similar to those in the lungs), from which run ducts, like the branches of a tree, all joining to go to the trunk. In this case the trunk is the gland cistern. Below this is another gland within the body of the teat (the teat cistern). The teat ends in two small holes, guarded by sphincter muscles, through which the milk is released.

Germs can enter the udder and produce an inflamed condition known as mastitis. The udder then becomes hard and tender to the touch, and swollen lymph ducts will show along the belly. This condition needs veterinary attention.

Urinary System

The urinary system (see Fig. 9.7) is concerned with removing water and unwanted substances from the blood by filtering all blood through the kidneys. It supplements the work of the lungs, skin and bowels.

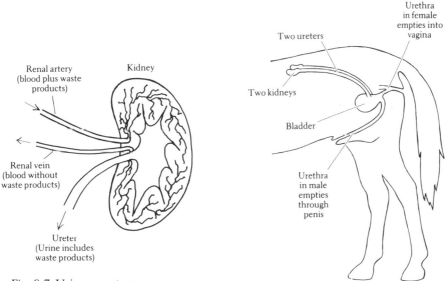

Fig. 9.7 Urinary system.

The kidneys maintain constancy in the horse's internal environment. They regulate water balance, acidity and alkalinity (pH), osmotic pressure, electrolyte levels, etc. The kidneys are affected by the composition of the blood, blood pressure, hormones, stress and drugs. There are two kidneys, each weighing about 700 g (23 oz.), located high in the abdominal cavity. Each consists of an outer cortex, in which waste products pass from the blood into collecting tubes. These form an inner medulla and empty into a central pelvis. Inflammation of the kidney is called nephritis or, if only the central pelvis is involved, pyelitis. Urinary calculi or stones may occur as a result of salts crystallising in the urine, but this condition is rare. Urine flows from each kidney down the ureters into the bladder. Cystitis is inflammation of the bladder. The exit to the bladder is controlled by a sphincter muscle which, when released, allows the urine to pass down the urethra to be discharged via the penis or vulva. A horse may pass up to 10 litres (18 pints) of urine per day.

Part III

Horsemastership and Stable Management

10 The horse in the stable

Housing

Although horses are hardy creatures and may be kept at grass all year round, it is often desirable that they should be housed in stables. Stabling provides protection for the horse and convenience for its owner. A horse that is fit has lost its protective fat; a groomed horse no longer has the natural protective oils in its coat. Such horses need protection from the elements. Where the horse is clipped out, as is the case with many working horses, it has lost its coat. Stables protect the horse from the cold, wet and wind during the winter months, and from heat, flies and sun during the summer.

From the owner's point of view, having a stabled horse is a convenience. The horse is at hand, is clean and dry, and is easier to feed and water. Even where adequate pasture is available, stabling the horse saves the grass. Indeed, it is possible to keep a horse without pasture at all, provided it is adequately housed. This is done at some racing stables by choice, and of necessity by some horse-owners in towns. Stabling also provides security and safety. This is especially so where the stables are near the house.

There are many other advantages of stabling. Obviously, it is easier to monitor and control the horse's food and water intake when it is inside. The stabled horse is easier to control, both as regards exercise and, where necessary, restraint. Stabling is, indeed, essential in cases of ill-health or sickness when isolation is desirable.

Requirements of a Stable

Sound stabling is a good investment from every point of view. Stabling need not necessarily be grand but ideally it should be purpose built. Although this represents a substantial capital investment, in most cases a stable block adds to the value of the property. Brick- or block-built stables are the best, but are very expensive, and most private owners today are content with stabling of the sectional wooden type. This is available from a number of manufacturers and varies considerably in both quality and price.

Whatever type of stabling is used, there are certain essentials to be borne in mind. Stables should be warm and dry and have adequate drainage. They should be free from draughts and yet enable the occupant to enjoy adequate fresh air. Wood, brick and blocks create a better environment than galvanised iron sheets or asbestos. Galvanised iron in particular has no insulation value and buildings constructed wholly or partly of this material are prone to condensation and overheating. This can be overcome to some extent by providing adequate insulation and boarding over the interior of the stable.

Good light is also an essential and this means providing electric lighting. In the stable itself, provision must be made for water and feed and, from the owner's point of view, stabling should be arranged so as to minimise on labour. Finally, but very importantly, stabling should be *safe*. Fire is a constant danger, but equally, electric wiring should be protected from inquisitive occupants, and there should be no projections that could cause injury.

Siting the Stables

The ideal site probably does not exist. However, the site should be level and well drained and, if starting from scratch, a concrete base should be installed with a good drainage system. Building controls require proper foundations, and in the case of new buildings, planning permission is needed. Approval by the local authority under the Building Regulations is always necessary. The stables should be protected from the prevailing wind, particularly from the north and east. Too many surrounding trees and buildings can prevent the free circulation of air, which is essential to health. If the stabling is to be erected near a dwelling house, the stable block should be sited downwind of the house. Consideration must also be given to ease of access, not only for people but also for routine and emergency vehicles.

The Stable Block

Today, most private owners prefer loose boxes as opposed to stalls where the horse is tied up. However, if converting an existing range of outbuildings not designed as stabling, the arrangement may dictate that the buildings are better converted to stalls.

The stable block itself, large or small, consists not only of loose boxes or stalls, but also of ancillary accommodation. Provision must be made for the storage of feed in the form of concentrates, and of hay and straw or alternative litter. A secure tack room is needed, and there must be somewhere to dry rugs and so on.

In a commercial stable there are additional requirements, e.g. an office and possibly a staff restroom or lounge. In every case, various items need to be stored: tools, wheelbarrows, horse-box or trailer, etc., and there must always be somewhere to dispose of manure. A manure bunker is easily constructed.

A variety of housing systems, modern and traditional, are in use. Stalls are traditional, but still have their place; however, in modern practice, loose boxes with ancillary accommodation are the answer.

Choosing a Loose Box

The first requirement for a loose box is that there should be adequate headroom: there should be a minimum of 3 m (10 ft). The recommended dimensions for a loose box are 3.7 m × 4.3 m (12 ft × 14 ft) for a large hunter, down to 3.7 m × 3.0 m (12 ft × 10 ft) for a pony. Stable doors should be 1.2 m (4 ft) wide and a minimum of 2.1 m (7 ft) high for a horse, and must open outwards or sideways. Latches and other fittings should not project on the edge of the door when it is open; proper stable bolts should be used, together with a 'kick catch' at the bottom. The stable door should be divided into two, the top part being hinged outwards and left fastened against the wall so that the horse can look out. The bottom half of the door should be about 1.4 m (4 ft 6 in) high for horses, with a metal covering along the top edge to prevent the horse from chewing the wood.

Foaling boxes, where required, must be larger: 4.6 m (15 ft) square is ideal. In every case, the box and its roof covering should be designed to keep temperature below 15° C (60°F), even in the hottest weather. Horses tolerate cold, but high temperatures cause them distress. In every case the box should give not less than 42 m³ (1500 ft³) space per horse.

Where stalls are used, a good size is 1.8 m (6 ft) wide by 2.7 m (9 ft) in depth, with a passage behind of a minimum width of 1.8 m (6 ft). Dividing partitions should be 2 m (6 ft 6 in) high at the front, and 1.5 m (5 ft) high at the rear. The aim should be to provide a minimum of 28 m³ (1000 ft³) of air for each horse within the building.

The floor should be non-slip and hard-wearing and should not strike cold; provision must be made for drainage. Concrete is the most commonly used material today, but possible alternatives are brick, tarmac, chalk or slats. Floors should slope from front to rear or vice versa so as to provide adequate drainage. Slopes are usually 1 in 60 in the loose box and 1 in 40 in any gutters. There should not be an open drain within the loose box, and thus there should be an open, trapped drain at the front or back.

Ventilation is of vital importance. Its object is to change the air frequently enough to keep it pure without causing draughts. Horses keep fitter and get less coughs when they have adequate fresh air. Warmth in the stable can be provided by clothing, and the important basic principle is to keep out draughts. Many people only close the top door to keep out driving rain, sleet or snow.

Each loose box should have an opening window, which should be protected by slats or mesh on the inside. The best form of window ventilation is the 'Sheringham system'. This directs cold air upwards so that it mixes with the warm air in the stable. Stables should have an opening in the ridge to allow warm stable air to escape.

Artificial lighting in each box is important as there must be adequate light to work both early and late during the winter months. Electric light fittings should be protected by means of a wire grille, or else be of the self-contained type. All fittings should be tamper-proof, and switches preferably should be outside the box, as should any power points.

The walls of the box should have kicking boards up to 1.2 m (4 ft) high and all internal walls should be free from projections.

Considerable care and thought is necessary over stable fittings: they must be accident proof. Tie rings must be well secured. Automatic watering systems are costly but labour saving. Plastic buckets are an adequate alternative, provided they are solid enough and have a capacity of at least 9 litres (2 gal). They should be placed on the floor in the corner near the door.

Hay may be fed in a deep manger, or from a rack or hay net fitted at eye level, although many people prefer to feed hay on the floor. Where a manger is fitted for feeding concentrates, it is typically placed in the corner at a height of approximately 1.1 m (3.7 ft).

Fig. 10.1 Modern labour-saving stables with provision to feed hay and concentrates from the outside.

Bedding

The type of bedding used depends on a number of factors, but an adequate bed is always essential. Bedding should provide comfort for the horse and insulate against damp and cold as well as helping to keep out draughts. Good bedding helps to keep the horse's feet in good order (with less wear on shoes), prevents injury and avoids the danger of the horse getting cast. It must provide comfort.

The bedding material used is a matter for individual choice. Availability, ease of use, and cost are all considerations. There are several sorts of bedding material.

Wheat straw is traditional and is the best drainage bedding.

Where used, it should be clean, bright and dust-free. In fact, it does not absorb liquid well, but it allows liquid to pass through to the drains below.

Barley straw and oat straw are slightly softer than wheat straw and do not remain as springy and free draining. They are likely to be eaten by the horse and, if used, it is vital to ensure freedom from dust. All bedding must be dust-free, but especially so in the case of barley or oat straw. Much modern straw contains a good deal of dust and short stalks.

Modern alternatives to straw are shavings and paper. Wood shavings are the easiest bedding to manage. They are absorbent and will not normally be eaten. Disposal is by burning. Shavings are quite economical and are easy to keep clean with regular skipping-out. Paper is a dust-free and absorbent bedding. It comes in two forms: shredded into long strips and diced into small pieces. Beds made of paper are less easy to manage than straw or shavings. Used paper bedding is burned.

Peat and sawdust are alternative bedding materials. Peat is expensive and is not easy to manage. Sawdust, though cheap, soon becomes hot and damp. Bark can also be used, but tends to be too damp, and the majority of people opt for one of the other alternatives.

In use, the bedding should be of a good depth and should be banked up against the walls of the box as a precaution against injury or draught. The box should be 'mucked out' thoroughly every day, all soiled and wet portions being discarded. If the horse is to be kept in, regular skipping-out of the droppings is essential.

Another way to manage a straw bedding is by the deep litter system. The dung and soiled portions of bedding are removed daily and fresh straw is then added. The whole bed is removed only when it is more than about half a metre (2 ft) thick. This method is labour saving in the short term and provides a deep and warm bed for the horse.

Clipping

Horses with thick coats get hot when working. If a horse overheats, it becomes distressed and this limits its performance. In order to try to keep cool, the horse sweats copiously, which leads to a reduction in body weight. Stabled horses in work are therefore clipped out. Clipped horses are easier to groom and to dry; they are less liable to chapped skin and to chills. Clipping also improves the horse's appearance and makes it easier to control external parasites and skin diseases.

A horse should not be clipped out until its winter coat is set, unless its performance is being limited. Usually, the first clip is given in October. Clipping is then repeated through the winter as often as necessary, usually at intervals of three weeks or a month. Ideally, the last clip should be given no later than January so that the horse's summer coat is not affected. However, it is always necessary to balance beauty with performance.

There are various types of clip, and which one is used depends upon the horse's lifestyle as well as fashion and the owner's preference. In the *full clip* the whole of the coat is removed. A variant is the *hunter clip*, where all of the coat is removed except that the hair is left on the legs as far as the elbows and thighs and a saddle patch is left on the back. In the *blanket clip*, hair is removed from the neck and belly only. *The trace clip* is a compromise commonly used on horses kept at grass: hair is removed from the belly as far as the traces, and also from the legs to a point half way down the forearm and thighs.

Even where the whole coat is removed, the ears are never clipped on the inside and the whiskers are left on all horses that live out. Heels and fetlocks can always be left at medium length and this is achieved by trimming with comb and scissors.

Clipping a horse properly is an art. It is also a lengthy process and the first essential is a good pair of electric clippers with sharp blades. The horse's coat should be cleaned thoroughly. The procedure for clipping is as follows:

(1) Set up the area to be free from obstruction and disturbances.

(2) Operate the machine from a proper power point, not from a light, etc.

(3) Rubber-soled footwear must be used in order to protect the operator from electric shocks.

(4) Adjust the blade tension screw with care. If too tight, it strains the motor; if too slack, it will not cut properly. Tighten until motor slows, then go back half a turn.

(5) Make sure the blades fit together as a pair and that they are sharp. They need sharpening regularly: blunt blades

do not cut properly and they strain the motor and pull the hair.

(6)   Use light machine oil or special spray to keep the blades well lubricated. The rotary bearing and the bearing surfaces in the clipper head should also be lubricated.

(7)   During clipping, the machine should be switched off at the power point from time to time and the blades should be allowed to cool and brushed clean. They must be kept free of dirt and grease; if necessary a brush dipped in surgical spirit may be used. The blades must be wiped dry *and re-lubricated* before being used again.

(8)   Start on the shoulder. Do not press but keep the points in contact with the skin, moving slowly in long strokes.

(9)   Work against or at right angles to the lie of the hair.

(10)   Keep the horse relaxed but use suitable restraints and distractions (e.g. a hay net) to maintain safety.

(11)   Keep the horse warm throughout; if necessary lie a blanket over its back and turn this forwards and backwards as in quartering during grooming (see next section).

(12)   Use chalk to mark the clip lines and use the actual saddle when marking the saddle patch.

(13)   When leaving on the hair on the legs, draw a line from a hand's breadth above the hock to the point of stifle, and from a hand's breadth below the elbow forwards and upwards to follow the muscle formation.

(14)   When clipping the head, slip the head collar on to the neck and make sure the blades are not over warm.

(15)   Long hair may be clipped with the lie of the hair to take off the excess and then back the other way.

(16)   After clipping, remove the blades, clean them thoroughly, lightly grease with petroleum jelly and pack away. Similarly, clean the head. Brush clean the air inlet and outlet on the body of the machine.

Grooming

The objects of grooming include the following:

To promote health. The process replaces rolling, removes waste products, e.g. sweat, keeps the pores open, and helps blood and lymph circulation.

To prevent disease. Parasites feed on dead hair and skin.

To improve condition. Massage is necessary for muscle and skin tone.

To improve appearance.

To ensure cleanliness. Grooming helps to keep tack and clothing clean and prevents sores.

Daily grooming also helps to improve the relationship between the horse and its owner. It is relaxing for the horse, but, if done properly, it is hard work for the groom. Grooming enables the owner to inspect and observe the horse closely.

Quartering

Quartering is done in the morning before exercise. After tying up the horse, its feet are picked out. Its eyes, nostrils and dock are sponged. The rugs are then thrown up or turned back, the roller being left in place unless the rugs are to be changed. The exposed parts of the body are then brushed down, a water brush being used for stable stains. If necessary, the mane is laid. The object of the exercise is to make the horse tidy.

Strapping

Strapping is a full grooming of a dry horse, given after exercise. The skin will have been warmed and loosened by the exercise, and the horse is relaxed. The procedure is as follows:

(1) Secure horse, rugs off, jacket off.
(2) Pick out feet into skip.
(3) Brush off mud and sweat with a dandy brush or a rubber curry comb.
(4) Body brush and curry comb:
    (a) near side, brush in left hand;
    (b) mane: brush and fingers (throw mane over far side of neck, brush crest thoroughly, and work mane back, dealing with only a few locks of hair at a time);
    (c) off side, brush in right hand;
    (d) head (head-stall around neck and steady horse's head with your hand);
    (e) tail, brush and fingers.
(5) Wisp or rubber, damp:

20 strokes on left neck – bang and slide in the direction of
the lay of the coat
20 strokes on left shoulder
20 strokes on left back muscles (gently on the loins)
20 strokes on left quarters
Repeat on other side; increase number of strokes as horse
gets fitter.

(6) Sponge eyes, nostrils and lips, then dock, sheath/udder.
(7) Water brush: lay the mane and tail, wash feet.
(8) Damp stable rubber to polish the horse.
(9) Brush oil on walls of feet.
(10) Bandage tail. Shake out rugs. Rug up and release.

Brush Over

A brush over or set fair is a light brushing when the rugs are
changed or at the end of the day. The object is to leave the horse
both comfortable and tidy. The feet are always picked out.

Washing

Sometimes it is necessary to wash the horse. Washing aids
grooming by clearing the pores of dust and grease. It may be done
because the horse is dirty or sweaty, or for appearance as when
showing.

If the horse is sweaty and blowing, it should be walked round
with an anti-sweat sheet topped with a cotton sheet until it stops
blowing. The sweat is then washed off quickly, the moisture being
removed with a sweat scraper. The horse is then rugged up and
walked dry.

When the horse is dirty, the basic procedure is the same, but in
this case, after cleaning off the dirt, particular attention should be
paid to the heels, which should be dried carefully with an old
towel. Cold water and a sponge should always be used; hot water
would open up the pores and the use of a stiff brush could aid
entry of germs into the skin.

Where washing is used as an aid to appearance or grooming, the
procedure may be summarised in this way:

(1) Wash out of the wind on a draining surface.
(2) Soak the horse's body, legs, neck and mane with hose and
sponge.
(3) Dissolve shampoo in warm water and scrub it into the coat

with a water brush on upper body, neck and mane and with a sponge on the legs, belly and dock.

(4) Wash the tail, putting it in the bucket and rubbing it between the hands.
(5) Wash the head and forelock with the sponge, taking care to keep soap out of the eyes. (Possibly use baby shampoo for the head.)
(6) Rinse the horse with the hose and sponge.
(7) Use the sweat scraper – metal side on body, rubber side on legs.
(8) Dry ears, legs and heels with an old towel, and the tail by swirling it.
(9) Lunge on clean grass to dry off, or dry under a heat lamp.

Clothing

In its natural state a horse needs no clothing other than that provided by nature. Horses in work need various articles of clothing for a variety of reasons.

A clipped horse in a field must be kept dry. This is achieved by using a New Zealand rug, of which several designs are available. Waterproof sheets and hoods may be used when riding out on exercise.

The horse also needs to be kept warm, e.g. after clipping or to avoid excess food consumption in cold weather. At night, the stabled horse will need an undersheet, a stable rug and a night rug, kept in place by a roller, and sometimes stable bandages as well. By day, an undersheet and day rug kept in place by a roller may be used, and sometimes the tail is bandaged. At exercise, the clipped horse may need a quarter sheet.

In order to help a horse to get dry without its catching a chill, an anti-sweat sheet is commonly used, with a lightweight sheet over it. The horse may also be 'thatched' under an anti-sweat sheet by putting a loose layer of straw on its back.

Another reason for the use of clothing is to care for the horse's coat by keeping it clean and sleek. Clothing will also help to deter hair growth during winter and assist when the coat is changing in the spring.

The horse out at grass may be troubled by flies. Protection may then be given in the form of a summer sheet and, in extreme cases, by means of a fly fringe, ear caps and veils.

Support bandages and exercise bandages provide support for the horse. Leg circulation may be assisted by the use of stable bandages. Sausage boots or poultice boots may be needed in certain medical conditions.

Protection may be required when the horse is travelling or even when it is at exercise. The normal clothing for travel is: head collar plus poll guard, sweat rug, jute rug turned back off the shoulder, roller, travel bandages or protectors, over-reach boots, tail bandage and guard, knee caps and hock boots. At exercise, skeleton knee caps, brushing boots or Yorkshire boots may be used, as well as a numnah or saddle cloth.

Feeding

There are ten basic rules of feeding, which always hold true:

(1) Make sure that thirst is quenched before feeding.
(2) Feed at regular times.
(3) Make no sudden changes in feed or routines.
(4) Use good quality feed that has been carefully stored.
(5) Base the rations on bulky feed such as grass or hay (forage). At least 25% of the diet should be forage, the exact proportions varying according to work.
(6) Keep concentrated feeds small: to feed more, feed more often.
(7) Keep work ahead of feeds (if the horse is having the day off tomorrow, feed less concentrates today). Work and feed should balance.
(8) Stabled horses need extra green food, e.g. a run in the field, or roots, apples, cut grass.
(9) Treat each horse as an individual. 'The eye of the master maketh the horse fat.' Pander a little to its preferences.
(10) Avoid work on a full stomach. At least an hour should be allowed between feeding and exercise.

These guidelines and some common sense will be adequate for most needs. However, a greater understanding of the principles is beneficial. Nutrition for horses is different to that of most other stock, and there is little agreement between authorities. Some of the research has been carried out with flat-race horses, which (because they are working and yet are immature) have special problems. Sufficient research has been undertaken to establish a reasonable pattern. What follows is not intended to suggest that the science of feeding should replace the art of feeding. If best results are to be obtained, art and science must be combined.

Nutrition

Nature's perfection is clearly seen in the relationship between animals and plants. Plants take up water from the ground together with carbon dioxide from the air, and, with the aid of energy from the sun, combine these two simple products into a more complex energy-storage product called carbohydrate. The animal eats the plant, utilises the carbohydrate within the muscle, and the energy is released in the form of body heat and activity. Carbohydrates include sugars, starches and cellulose (fibre). In digestion, carbohydrates are broken down to simple sugars to go through the gut wall and are then built up again within the horse into more complex sugars and animal starch (glycogen). The glycogen is stored mainly in the liver as a readily available source of energy.

If longer-term storage is required, carbohydrates can be built up by both plants and animals into lipids, which exist in solid form as fats, or liquid form as oils. Typically 17% of a horse's body is made up of lipids, but the horse does not require a lot of lipids in its diet, except for endurance work.

Plants also take up elements, notably nitrogen, to form (in combination with carbon, hydrogen and oxygen) proteins. Every protein is made up of parts like building blocks, called amino acids. In digestion the plant proteins are taken apart into their amino acids to pass through the gut wall. They are then recombined to form different animal proteins. About half of the amino acids can be made up or synthesised within the horse's body, and therefore do not need to be in its food. The remainder (called essential amino acids) must be in the food. In practice the mature grazing horse or the stabled horse fed on reasonable hay plus some grain or grain product will usually get an adequate mix providing it is not stressed by strenuous demands. However, in all probability a chemical analysis would show that this diet was low in one of the essential amino acids, namely lysine.

Our grandfathers overcame this problem unconsciously by including some peas or beans in the feed. Modern food compounders do the same thing, using either soya-bean meal, which is rather high in protein, milk powder or lysine.

Protein that is fed surplus to requirements has the nitrogen removed by the kidneys. It is then stored as lipids or used as a source of energy. In the horse, the protein is used to form the body tissues, first in growth, and later to keep up a continuous repair and replacement service to all parts of the body.

In fast work there is more wear and tear on the system. Slightly more protein is then needed. Racehorse cubes contain more protein than standard cubes but are designed for growing two- to three-year-old racehorses.

Protein is also needed for reproduction, especially during the last three months of pregnancy and during lactation (milk production).

The best reserves of energy in plants are in the seed or grain. Thus, cereals are used in horse feed for extra energy. Protein can be obtained from animal sources, e.g. the fish meal and bone meal used in some horse feed cubes. It may also be obtained from plants of which the pulses (peas and beans) provide concentrated sources. Dried milk powder is another source of concentrates. However, the concentration of protein in cereals is usually more than adequate for the horse's nutritional needs.

In addition to carbohydrate and protein, horses need minerals, vitamins and water. Plants, particularly the deep rooted ones, take up minerals from the ground. In general, they provide an adequate mix to meet the horse's requirements for building bone and for the proper functioning of the various systems.

Earlier this century there was less intensive production from the land than is the case today. Fertilisers, weedkillers and highly productive varieties of plants were then in their infancy. Until quite recently, crops contained more weeds and herbs than modern crops. Weeds and herbs have a good mineral balance, whereas modern, fast-grown, clean crops require supplementing by minerals when fed in some circumstances, for example in the case of young stock.

In general, the stable-fed horse needs extra salt. It is also important that the balance of the minerals (notably calcium and phosphorus) is kept at a reasonable level or weak bones will result. For example, bran is very bad in this respect. It is too high in phosphorus and also contains a substance that inhibits calcium uptake. Bran needs balancing with ground-up limestone to provide extra calcium.

Plants also contain some of the vitamins that the horse needs in order to function properly. The horse makes some vitamins within its body with the aid of sunlight and the bacteria in the hind part of the gut.

Feeding Practice

The horse prefers a diet based on roughage and one that is free from moulds or taint. Its digestive system is designed for a small intake at frequent intervals. If the horse is fed two large feeds a day, the food will be less well utilised and therefore money is wasted. Moreover, although the horse appreciates some variety, it does not like sudden change, particularly of the roughage part of its diet. This is because the roughage is broken down in the horse by bacteria which develop to meet a particular need. If there is a major change in diet, the bacterial population will change, but this takes time.

Within the gut, cereal-based foods tend to form a mass like bread dough, which the digestive juices have difficulty in penetrating. Digestion is more efficient if cereals are mixed with roughage. The inclusion of a little bran, dried grass, nuts or soaked sugar beet with the cereals can aid digestion. Chaff (chopped hay or straw) may also be included. It is a useful aid to digestion to offer the stabled horse a small amount of hay before the morning corn feed. The hay will go into the stomach ahead of the corn and help to break up the 'dough'.

Oat and barley straw are as nutritious as low-quality hay. Straw can therefore usefully supplement the diet of ponies wintering outside. However, neither hay nor straw supplies an adequate diet without supplement: feed blocks are useful in this respect.

Concentrate feeds based on corn are important highlights of the day for the stabled horse, and are best given at regular times. The horse is, of course, capable of being flexible in feeding arrangements when away from home. Because the horse has a keen sense of smell, all feed containers must be kept scrupulously clean and mangers must be checked before the next feed is put in.

Water

Between 50 and 70% of the horse's body weight consists of water, the actual percentage depending on age and condition. Water has a number of essential functions:

it assists in the maintenance of a uniform body temperature and in the removal of surplus heat as sweat;

it acts as a medium in which essential chemical reactions take place;

it acts as a solvent in which substances can be dissolved and transported in the body;

it helps give shape to the body cells;

it forms the base of urine for excreting waste;

it forms the base of milk for lactating mares.

In the wild, animals tend to drink at dawn and dusk, but stabled horses need more water because they are eating dry feed. A typical quantity for a stabled middleweight hunter would be 37 litres (8 gal) per day. However, the quantity varies according to the animal, the work, the weather and the food. Water should be free from taint, poisons and disease. At competitions it is best to use a tap or take one's own water rather than allow the horse to drink from a public trough where germs could be picked up.

In the stable yard, if buckets are filled from a trough, there should be a dipper bucket on a hook so as to avoid the water in the trough becoming contaminated by the dirty bottoms of the stable buckets. All troughs and drinkers must be kept clean and should be emptied and scrubbed regularly. Often the water in a drinking bucket in the stable will be rejected because it has got stale or fouled. Water is therefore best changed rather than topped up.

Except when a horse comes in hot and tired, it is usual to give the animal free access to water at all times. If a water bucket in a stable is empty, it must be refilled several minutes before the horse is offered a corn feed. Heavy drinking after a meal could wash food through the stomach and cause digestive upsets. However, water is heavy and can fill the stomach when it is required to be empty. For this reason water should be removed from the stable two hours before important fast work, but not before endurance work when the horse might get dehydrated.

The hot, tired horse may be offered a little 'chilled' water, which is water with the chill removed, although cold water is adequate.

After long periods of heavy or fast work the horse may be dehydrated, particularly in hot weather. In these circumstances it should be offered only a limited quantity of water every ten minutes until it has satisfied its thirst. Essential minerals for adding to drinking water are called electrolytes, and when horses are

dehydrated they should be offered both plain water and electrolytes simultaneously so that they can choose according to their
need. Horses dehydrate after any period of heavy sweating and on
long journeys on hot days, even though, with modern rugs, this
may not be apparent as sweat.

Minerals

Mineral balance is important. Even if all the required minerals
were present in a horse's diet, the horse might not extract or utilise
them properly. Mineral nutrition therefore contains a degree of
mystery on which the supplement salesmen prosper!

Calcium (Ca) is needed for bone formation and also for nerve
function. The horse is unable to utilise all of the calcium in its diet.
This mineral is found in grass. The stabled horse's diet may need
to be supplemented with bone meal or ground chalk or limestone
as the balance of calcium to phosphorus should always favour
calcium. Good hay has about twice as much calcium as phosphorus. Cereals have more phosphorus than calcium. Vitamin D is
necessary for the absorption of calcium. The chief source of this
vitamin is the action of sunlight on the skin. Thus overwintering a
young horse indoors without a source of vitamin D and feeding it
large quantities of bran would be a recipe for disaster.

The phosphorus (P) needed for bone formation is generally
supplied by grass and grain. The sodium (Na) and chlorine (Cl)
required by the horse are best supplemented for the stabled horse
by the addition of a little salt (NaCl) in every feed and a salt lick to
top up as required. A deficiency of salt leads to early fatigue and
so the horse should be given 25–100 g (1–4 oz.) a day according to
work and the weather.

The horse's food usually contains sufficient magnesium,
manganese, copper, fluorine, iodine and cobalt for its needs.
However, in some circumstances the diet may be short of
potassium, iron, zinc, sulphur and selenium. Some geographical
areas have deficiency problems, and hay or cereals grown there
may need supplementing. Other areas have problems of excess,
whether in the soil itself or as a result of pollution.

An iodine salt lick is said to help horses that are prone to sweet
itch.

The amount of iron in the diet may be adequate until the horse
is stressed by high performance, when double the quantity is
needed. A high-performance horse showing slight lack of form

should have its blood checked for anaemia.

If good-quality and carefully stored foods are used, most horses will need few supplements unless they are stressed for growth, breeding or high performance. A little deep-rooting Russian comfrey cut each day will help the stabled horse in summer, and a single supplement (not a home-brew cocktail) may help in winter.

Vitamins

Vitamins are a complex group of organic substances that are essential, in small quantities, for the normal function of metabolism in the body. Usually, they cannot be synthesised in the body but they occur naturally in certain foods. The more important vitamins are as summarised in Table 10.1.

Table 10.1　Vitamins.

Name	Notes
A　(retinol)	Fat soluble
B　complex 　　B_1 (thiamin) 　　B_2 (riboflavin) 　　B_6 (pyridoxine) 　　B_{12} (cobalamin) 　　niacin (nicotinic acid) 　　pantothenic acid 　　folic acid 　　biotin 　　choline	Water soluble Used to be called vitamin G Used to be called vitamin H
C　(ascorbic acid)	Water soluble
D　(calciferol)	Fat soluble
E　(tocopherol)	Fat soluble
K　(menaquinone)	Fat soluble

A (Retinol)

Fresh plant material contains ample carotene which is converted into vitamin A in the horse. This is a particular reason why stabled horses need green food, carrots or cod-liver oil. Lack of vitamin A causes poor feet and coat, weepy eyes, restlessness and reduced resistance to infection. Breeding mares need high levels of vitamin A and carotene.

B Complex of Vitamins

These are water-soluble vitamins and are not stored by the body in significant amounts. Typical of the complexity of vitamin nutrition is B_{12}. This essential vitamin can be synthesised by the horse (providing it is not short of cobalt) but parasites either use up or prevent its uptake. Racehorse trainers who find highly trained horses going off their feed resort to an injection of vitamin B_{12} to 'perk up' the appetite for a few days prior to a race. There is evidence that it helps the red blood cells.

Good-quality feed, and production by the bacteria in the gut, normally supply adequate B_1 (thiamin), B_2 (riboflavin), B_6 (pyridoxine) and B_{12} (cobalamin). However, folic acid (a deficiency of which causes anaemia) may sometimes be in short supply, particularly to meet the needs of growing stock. Biotin and choline are also needed.

C (Ascorbic Acid)

This is essential, and is sufficiently supplied by the bacteria in the gut. Like the B complex, vitamin C is water soluble; other vitamins are fat soluble and are stored to some extent in the body.

D (Calciferol)

The horse forms its supply of vitamin D with the aid of sunlight. However, the horse has a limited storage capacity, and although good quality hay will in general provide sufficient vitamin D, a horse which is kept inside all winter may need a supplement, e.g. cod liver oil.

E (Tocopherol)

This vitamin is found in fresh foods. It does not store well. Mares being brought on to breed early will need extra vitamin E, and stallions will benefit from being given this vitamin early in the season. Adequate vitamin E is essential for stamina and high performance. It also reduces nervousness in some cases and this may also be related to its usefulness as a treatment for the 'tying-up' syndrome (azoturia: see chapter 5). Vitamin E allows the body to use oxygen more efficiently, but it can be properly utilised by the horse only if the diet has adequate selenium; this mineral is needed to form the enzyme which uses the vitamin. In areas where there is a shortage of selenium, this should be included (in minute quantity) in the daily diet of performance horses.

K Group

This is essential for the normal clotting of the blood. It is present in green food and is made by bacteria in the digestive tract.

Energy

Carbohydrates are the best source of energy for horses. To work out how much energy the horse needs should be relatively simple. The more work the horse is required to do, the more energy must be given. The difficult part is to measure the work. If the horse is given more energy than it can use, the surplus will be laid down as fat. On the other hand, if the horse uses more energy than is provided, reserves will be drawn from the body and the horse will grow thinner.

In order to apply this principle to equine nutrition, it is necessary to measure the energy in food. Calories could be used, as in human nutrition, but the common convention is to express the energy as so many joules in a kilogram of food. A joule is a measurement of energy (as is a calorie) and because there will be thousands of joules involved, measurement is by a multiple factor called megajoules (MJ).

All of the food's energy potential cannot be extracted by the horse. For example, the horse can digest cellulose fibre but not lignin fibre. The information needed is therefore concerned only with the digestible energy (D.E.). Thus the best information about food for work is expressed as megajoules of D.E.

Protein

Protein is essential for all living organisms, and the horse owner needs to understand the protein that is required by the horse and should be provided by its food.

A method that easily measures food for protein content is to analyse the food for nitrogen. (Protein generally contains 16% nitrogen.) This measure expresses the percentage of crude protein in the food, but is only a rough guide. Not all of the crude protein is digestible, and so some feed tables show a figure for digestible crude protein (D.C.P.). Not all horse feed is sold with the protein percentage declared, and basic ingredients such as hay may be very variable in protein content. Generally, grass has higher protein content in the spring and when it has young shoots. The protein content decreases as the summer progresses and the grass flowers and makes seed heads. Thus, for example, vacuum-packed wilted grass may be higher in protein than most hay and this should be allowed for when feeding.

Rationing

When deciding on the feed for a horse, it is wise to bear in mind the theory of rationing. Horses are fed for maintenance, i.e. to maintain them in their present state. Their food provides energy for the muscles of the internal organs and for grazing, maintains body temperature and continuously replaces cells to keep the body in good order. Horses are also fed for production. This can be broken into different categories:

(a)   growth from the day the horse is born until it stops growing at 4 to 7 years old;
(b)   lactation of the brood mare from the day her foal is born until the day it is weaned;
(c)   growth of the embryo into a foal within its mother (most of this growth occurs in the final third of the pregnancy);
(d)   body repair, regrowth after major injury or disease;
(e)   fattening;
(f)   work (this can be broken down into light work, medium work, heavy work or fast work).

To feed for maintenance, the main criterion is the size of the animal: bigger animals need more food. In practice, maintenance for the horse in the field is produced by grass supplemented with hay or straw and possibly some hard feed in winter. Maintenance for the stabled horse comes from hay or other forms of conserved grass. If the horse requires a large quantity of food for production, the hay must be of high quality and therefore less bulk is required. There is a limit to the gut capacity of a horse, and if a performance horse is filled up with large quantities of low-quality maintenance food, it will have neither room for adequate production rations nor zest for its job.

To feed for production, the main criteria is the amount of production required. Thus, with the pregnant mare, the extra feed is increased gradually through the last third of the pregnancy. With the competition horse, the feed is increased as the horse becomes fitter and is able to work harder. Production rations are mostly based on cereals. Good grass or very high-quality hay will produce a little above maintenance and therefore provide some production.

The horse can eat, each day, hay and concentrates weighing

Fig. 10.2 Horses, like humans, need to watch their weight. Rationing and medication can be used more accurately if the horse's weight is known.

Fig. 10.3 An easy way to watch a horse's weight is by using a weigh-tape.

about $2\frac{1}{2}\%$ of its body weight (2% for ponies and up to 3% for young stock). The ratio of hay to concentrates will depend on hay quality and production required (see Table 10.2).

Table 10.2 Feeding the stabled horse.

	Hay (%)	Concentrates (%)
Maintenance	100	0
Light work	70	30
Medium work	50–60	50–40
Hard or fast work	25–30	75–70

The Appendix contains a system for estimating food requirements in greater detail.

11 The horse at grass

At one time, horses in England were only turned out to grass during their rest season. This is still the case with most hunters, which are kept at grass during the summer thus giving them a rest from work. However, many horses are kept wholly at grass throughout the year. This is a method practised by many private owners. Another possibility is the combined system, whereby the stabled horse spends part of each day at grass and the remainder in its stable.

Requirements of a Paddock

Horses need daily attention even when turned out at grass. From the owner's point of view easy access at all times of the year is important. A paddock near the house is ideal but not always possible. Adjacent activities should also be considered: a railway line may be tolerable but not, perhaps, a go-kart track.

Owners should also consider how the land lies: a level field is to be preferred to a steep one. Steep fields can create stress for exuberant Thoroughbreds and youngsters. They also limit the possibilities of ridden exercise within the field.

The aspect of the paddock is also significant. A field is warmer if it faces south, and the grass will grow earlier in the spring. Trees and hedges are a great asset, as they provide natural shelter from wind and rain as well as shade from the sun in the summer. If there is no natural shelter, an artificial field shelter or wind-break should be provided (see Fig. 11.1).

In winter-time, horses will tend to eat from the bark of trees, so ideally trees within reach should be protectively fenced.

Adequate fencing of the field boundaries is essential. Hazard areas, such as rabbit holes and obstructions, must also be fenced.

Fig. 11.1 An artificial wind-break built in an exposed field.

Large stones and rubbish such as old corrugated iron sheeting should be removed. The use of barbed wire should be avoided.

Soil type is of great significance. Light sandy soils stay dry all year round, but they grow poor grass, particularly in midsummer. However, they are best for riding on, and this may well be an important consideration. Heavy clay soils (even with drains) get deep and muddy in winter. They are slow to start grass growth in spring, but they are productive. A serious disadvantage is the tendency of such soils to go so hard in midsummer that they are a hazard to ride on. Loam, a mixture of sand, silt and clay, is a good compromise between the two extremes.

Paddocks should be properly drained. Drains are expensive to install and should be kept in good order. Ditches should be tended regularly and, if there is a drainage problem, expert advice should be sought.

The paddock must have an adequate water supply. This may be either natural or artificial. Stagnant ponds are best fenced off, and the ideal natural supply is an unpolluted running stream, preferably with a good stone bed. Sand or mud is easily stirred up by horses and can result in colic when swallowed over a long period.

Water troughs must be properly sited and adequately protected. They should be of good construction and preferably purpose made. A trough which has a sharp edge at knee height, such as an old bath, can injure horses, and any ball-cock or other projections should be well protected from inquisitive horses. Pipes rising from the ground should be protected against frost.

A trough sticking out from a fence is not really ideal. The ground around the trough will tend to get poached in winter, and hence it is best to excavate the top soil, lay a builder's permeable membrane, and then cover the area with stone topped with sand. The same procedure should be followed in gateways and in front of field shelters if the field is to be used during the winter months.

Fig. 11.2 Water troughs need hard standing on clay soils. However, they are best recessed into the fence.

Good fencing is pleasing to the eye and a good investment, both financially and otherwise. It reduces worry about stock getting hurt or straying. Fencing must be safe and tidy, must stand up to pressure, must be easily maintained and not easily chewed, and must be fairly inexpensive. A good fence is plain high-tensile wire with strong straining posts, with the bottom strand of wire 0.3 m (1 ft) from ground level. The traditional post-and-rail fencing is excellent, though on the costly side. Well-creosoted rails are not usually chewed but tanalised ones are.

A possible compromise fence is to have a strand of taut plain wire along the top of the fence to prevent chewing, with a rail or two and a plain wire at the bottom. If cattle are also to be fenced against, the lower strand should run through insulators and be electrified. If this precaution is not taken, cattle may graze under the wire and tend to push the fence over.

Where sheep are a problem, a different type of fencing is needed: special sheep wire is available. An alternative is to have two lower strands of wire which must be kept taut for safety.

Several substitutes for wooden rails are now available in both metal and rubberised webbing. They are strong, durable and maintenance free, but must be erected with considerable tension to stay stock-proof and smart.

Where it is necessary to create a temporary partition in a field, as for strip grazing or to prevent horses coming right up to the permanent fence, an electric fence can be used. This should be of the type specially designed for horses, with a broad and easy-to-see band.

Paddock Management

The owner's aim is to provide nourishing food for as long as possible during the year. This requires a sward of good nutritious grasses and a base that will stand up to the wear and tear imposed by stock. A further consideration is to try to reduce reliance on drugs for controlling parasites. This can be achieved in three main ways: (a) the use of mixed grazing; (b) rotational grazing; and (c) good grass husbandry. In small paddocks near to home it is important to keep the field as free as possible of droppings, by collecting them on a regular basis before they have time to cause trouble.

The division of land so that good management can be practised raises problems of stocking rates and size of paddocks. The question of size includes so many variables that it is difficult to be specific. Any clear-cut answer is open to misinterpretation. However, assuming 16-hand horses weighing about 500 kg each, with well-managed paddocks on loam soil about one horse per acre (which is just over two per hectare) is about right. This stocking rate allows for some poaching of the ground in winter, with its inevitable effect on grass production in summer. Even without the use of fertilisers, there will be too much grass in summer unless some is conserved. If the best use is to be made of land, large fields will be split into paddocks. Hopefully, only one paddock will then become poached in winter.

During the summer the horses should be moved from paddock to paddock in rotation, with each paddock having at least three weeks' rest. When a paddock has been grazed and the horses have been moved on, fertiliser should be broadcast on it and the paddock should be rested. Horses should not be allowed to graze on a fertilised paddock where the granules can still be seen, as this would risk colic or poisoning through ingestion of the chemical.

Grass for horses should not be very lush or high in nitrogen content. If a high-N compound fertiliser is used, it should be at a lower dosage than that advised for cattle. A typical pattern might be a 25 : 10 : 10 analysis at 200 kg per hectare when the grass starts to grow in spring. Thereafter, just 100 kg per hectare of straight nitrogen fertiliser should be applied after each grazing, although considerably more should be applied before a hay cut. After a conservation cut, a fertiliser with extra potash is often used to make good the nutrients taken from the land.

Excess grass in summer calls either for extra stock (horses, cattle or sheep) or conservation. It can be sold to a farmer to take for hay or silage. There are problems in taking hay for one's own use unless the acreage is sufficient to justify owning the equipment. Agricultural contractors can be employed to take hay, but horse-owners are often their smallest and least important customers. Thus, the hay is not turned often enough or will be baled wet or old and the hay will be poor. Good hay can be made by hand, using old-fashioned tripods, if one has the time and energy.

The advantage of cattle and sheep is that they tidy up the pasture and consume the worms that damage horses but not themselves. Cattle and sheep need especially strong fencing to restrain them. If the extra stock are one's own, when winter comes some of them must be housed or sold; however, sheep do not poach land to the extent that horses or cattle do.

Where land is poached after the winter, a good harrowing aerates the soil and levels the surface for grazing. If a good riding surface is required, the harrowing should be followed by a ring- or flat-roll. The timing of this treatment is critical to the day.

As the summer progresses, the pasture should be mowed to top weeds and rejected grasses. This is especially desirable if only horses are kept. Mowing is important, because without it all the wrong plants will prosper. On very hot days, the field may be chain-harrowed to scatter horse droppings in the hope that the sun will dry out and kill both worm eggs and larvae.

Fig. 11.3 Mixed stocking can be beneficial financially and is good for the pasture.

Pasture Improvement

Poor pasture can always be improved. Pasture improvement is an aspect of horse management that is ignored by many people. The foundation of good pasture is the soil, and just as the gardener can improve his vegetable patch, so can the paddock-owner improve his soil. A sandy soil will benefit from organic manure; this helps to hold both nutrients and moisture. A clay soil will benefit most from drainage.

The acidity of the soil is important and should be considered every five years or so. The local lime company's representative will take soil samples and analyse them. Acidity is measured on a pH scale, with 7 being neutral. Lower numbers are acid and higher numbers alkaline. For grass a pH of 6.5 is about right. If the soil is too acid, the wrong plants prosper, and if it is too alkaline, the grass cannot take up certain minerals from the soil. This is bad for the nutrition of growing stock. If the soil is very alkaline, a good dressing of farmyard manure will tip the balance towards acidity. If possible, horse manure should be avoided as it may contain worm eggs.

Acid soil (pH 6 or lower) needs different treatment, and a good dressing of lime or ground chalk should be applied.

Apart from calcium (which is in lime and generally freely available in the soil) the main nutrients to consider are nitrogen (N), phosphorus (P_2O_5 is phosphate) and potassium (K_2O is potash). Nitrogen is leached out of the soil by rain and so is only used during the growing season. Phosphates and potash are needed for plant efficiency. Local agricultural merchants can put paddock owners in touch with a fertiliser representative who knows the soil types in the locality. If .there are likely to be phosphate or potash deficiencies, he will advise the appropriate compound fertiliser to use.

However, too much quick-acting phosphate will upset the horse's calcium : phosphorus balance. Some people are anxious about the use of 'chemical fertilisers' and prefer to use those with an organic base. These are generally slower acting and more expensive, but they provide some minor nutrients which the soil might lack and they may well be free of fluoride which is often found in compound fertilisers. Owners of breeding stock will be anxious that their mares cycle regularly and produce healthy foals every year. They should, therefore, consider soil nutrition as it affects horse nutrition.

Phosphate encourages clover which produces free nitrogen from nodules on its roots. A little clover is desirable in the sward; one plant of wild white clover for each square metre is sufficient. An excess of clover makes the herbage too rich for horses.

The bulk of the sward should come mainly from a late-flowering perennial ryegrass of prostrate growth. A grass called S23 meets this need. Creeping red fescue is productive and gives a good turf; crested dog's-tail also resists treading. In dryer areas, a little smooth-stalked meadow grass may be added to the mixture, and in wetter areas rough-stalked meadow grass is good. Cocksfoot is hard-wearing but grows into clumps. Timothy is persistent, and most horses prefer tall fescue.

The main ingredients in the grass mixture should be represented by two similar varieties. A suitable seed mixture for a hard-wearing, palatable and productive paddock might be as follows:

Species	kg/ha
Perennial ryegrass (two varieties)	18
Creeping red fescue	5

Crested dog's-tail	1
Rough- or smooth-stalked meadow grass	2
Cocksfoot, timothy, tall fescue (two of each)	6
Wild white clover	1
	___
	33 (30 lb/acre)

This mixture is mainly for grazing, but an occasional cut of hay may be taken from the sward. It is best not to put herbs into the mixture as they make it harder to use weedkillers (called herbicides!) and for hay to dry out. However, horses like herbs, the deep roots of which bring up minerals from the soil. A compromise can be achieved by hand-sowing a strip of herbs along the fence. These herbs should include chicory, ribwort, yarrow and burnet.

Horse pasture is rarely ploughed up, but the mixture suggested can be used to renovate old pastures as well as to create new ones. If a paddock is thin, renovate by mixing the seed with fertiliser, applying by means of a spinner or by hand-broadcasting. Seeding must be done in spring or autumn. Chain-harrow before seeding and ring-roll afterwards.

Topping and selective scything will control weeds. Ragwort needs pulling and burning, and docks, thistles and bracken can be spot-treated using a knapsack sprayer. Major infestations need specialist herbicides, and if necessary expert advice should be sought. Excess clover is reduced by taking a hay crop or by spraying with a herbicide that contains MCPA and is designed to kill broad-leaved plants.

Many fields are badly neglected but can be renovated without too much difficulty. A suggested programme might be: make good the fencing, water, drainage and lime. Put in a herd of cattle to eat it bare, harrow hard, fertilise, seed and roll. Make the early grazings light and avoid using cattle until the grass is well established. The routine then is horses, cattle, top if necessary, fertilise and rest. This cycle is then repeated. Grazing too tight (5 cm; 2 in) should be avoided, as should letting the grass get so long that treading wastes good food.

Working from Grass

Many advantages are claimed for grass-kept horses. It is a natural system, and has much to recommend it. Less straw and hay are used and less time is spent on routine management. The horses need not be ridden because they exercise themselves.

This system has corresponding disadvantages. The horse is often too fat in summer, and in winter it is often wet and muddy. Where a horse is kept at grass, it should be caught up and handled daily if possible. Supplementary feeding will be necessary in wintertime, and where there is insufficient grass during the summer as well.

It is easier to operate on the 'combined system' where the horse spends part of each day at grass and part in its stable. Thus, in summer the horse, or more particularly the pony, is shut in for part of the day to limit its food intake and also to protect it from flies. In winter the horse may be stabled at night or else shut in well in advance of being ridden so that it can be dried off. This is an ideal method for keeping hunters.

Horses working from grass in winter will usually be given a trace or blanket clip. This makes grooming easier and the horse can gallop about without undue sweating. A clipped horse will need a good New Zealand rug. The best designs have two straps at the front, a well-fitted back and good leg straps so that the rug will stay in place even when the horse rolls.

The unclipped horse can stay out without a rug, but its long coat means that there is the problem of drying its back in wet weather before saddling up. The best method is to put it in the stable with a thatch of straw, lightly covered with a cut-open lightweight sack held in place with a loosely fastened surcingle. After half an hour or so, the worst of the mud may be brushed off. It is undesirable to remove the natural protective grease from an unclipped horse living at grass, and so only a dandy brush should be used. If the back is not cleaned before saddling up, mud will be ground into both the saddle and the horse's back, which could cause sores.

A wet horse must not be left in the stable after riding. If it is to remain stabled, it should be thatched. If the horse is to be turned out, this should be done straightaway so that the horse can roll and keep on the move. A cold, draughty stable and a wet horse will soon lead to a chill.

A variant of the traditional system which, although expensive in initial outlay, offers many advantages of the paddock without the

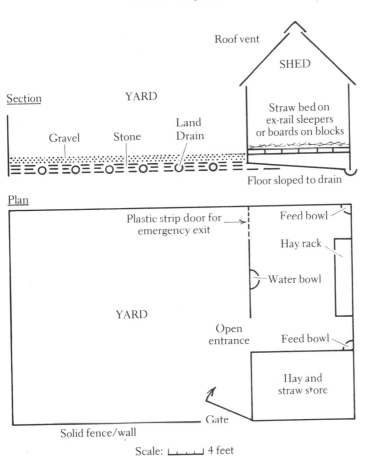

To house two ponies

Roof vent

SHED

Section   YARD

Straw bed on
ex-rail sleepers
or boards on blocks

Land
Gravel   Stone   Drain

Floor sloped to drain

Plan

Plastic strip door for → Feed bowl
emergency exit

Hay rack

Water bowl

YARD

Open
entrance   Feed bowl

Hay and
straw store

Gate
Solid fence/wall

Scale: ⌊___⌋ 4 feet

Fig. 11.4 Accommodation with exercise area.

mess, is that of the shelter/stable with a free-draining sand yard attached (see Fig. 11.4). The ever-increasing cost of bedding and labour means that this system will become more important.

Whether in field, yard or stable, there is always the problem of how best to feed hay and concentrates. In the field, concentrates may be fed in bowls. Those set in car tyres are good as they tip less easily and have fewer sharp edges. Bowls should be spaced out and it is a good idea to have one more bowl then the number of horses as there will be plenty of swapping when the greedy see off the timid. Hay can be fed on the ground along the fence line to reduce treading. On clay fields, the fence line soon becomes poached.

Nets are laborious and a slight hazard, and hay racks tend to be wasteful. Feeding hay in a shed may induce kicking or biting. If available, cheap plastic pods set along the fence are the best answer. Hay is pushed in at one end of a tube, which is like a flute, and the horses pull it out through the holes.

Management Routine

When horses are kept solely at grass, a daily visit is essential. The experienced eye will spot small details that may give a clue to more significant events.

Is the horse always standing away from hedges bordering roads because children throw stones at it? Is the good summer weather going to produce a heavy crop of acorns, which will fall and be of sufficient quantity to poison the horse if eaten? Is the coat 'staring', indicating an increased worm burden? These are the sort of questions to ask. Changes are sometimes so slight that only a visit with someone who has not seen the horse for several weeks may draw attention to subtle changes.

The daily check of the horse at grass can be made entirely by eye. The horse should be looked over for injuries. Particular attention should be paid to the feet. A hand run under the belly, down the legs and over the back may detect something which the eye has missed. This last check may need two people for safety and convenience: it may be carried out weekly. It is a good plan to have a hoof pick and to lift, inspect and pick out the feet at least once a week.

The daily check must include the water supply and the general state of the field, including the grass. By October, for example, grass has less feed value, and hay or concentrates may need to be started or increased.

The field boundary should be walked regularly and any fencing defects noted and made good. Areas by footpaths, public roads or people's gardens all need regular checks for items thrown away, particularly wire, tin, glass or rubber. Like litter collecting, weed removal needs constant attention: poisonous plants such as woody nightshade, foxglove and so on should be uprooted and removed. Ragwort repays pulling up. Docks and thistles do not spread if the field walker carries a sharp billhook or light scythe and constantly tops them.

Horse people have a bad name because of fields that are a mess of mud and weeds, fallen-down jumps, botched-up fencing and tatty buildings. This reputation could be scotched by careful appraisal by the field-owner, and by good husbandry and constant vigilance.

Getting Up from Grass

There are two methods by which the grass-kept horse may be changed over to a stable regime. The choice of method must depend on individual circumstances and facilities.

Gradual Method

This is the better method because it avoids sudden change, which is a shock to the horse's gut and to the horse itself. It is particularly suitable for those with a paddock close by the stables.

If the grass-kept horse is not getting any hard food, this is introduced at least a week before the horse comes in permanently. Hard food is introduced gradually, starting with a small feed and working up to two larger feeds each day. If there is an empty stable to hand, the feeds can be given in the stable. Keeping the horse stabled for a while reduces the time spent grazing and prepares the animal for its less tubby outline.

This is the time to check the teeth and have them rasped if necessary. The horse will also probably require shoeing, and it is also the best time to have the horse immunised against tetanus and equine influenza. After such inoculations, a horse needs several days without work to avoid sweating or blowing.

Walking exercise is often started from the field. This has the advantage of keeping the horse placid in its mind. A corn-fed stabled horse, particularly if freshly clipped, can be quite 'on its toes'!

The first few days a horse is fully stabled after coming up from grass are always critical from the point of view of health. The two danger areas are the digestive and respiratory tracts. The stable must be well ventilated and have no stagnant air. The only occasion a cigarette should be permitted in a stable is when blowing smoke into an empty box to test air movement. A bee-keeper's smoker will do a better job if one is to hand. The top door should be kept open at all times.

Hay must be well dunked in a trough and fed damp; it is not sufficient to hose it. 'Haylage' (trade names such as 'Hygrass', 'Horsage' and 'Propac') is a cross between hay and silage and can be fed with advantage at this time. It is expensive, but it is dust-free and highly nutritious.

Feeds must be laxative and slightly moistened. They should include some bran or chaff to help the bowels, which are accustomed to a high-fibre diet. Exercise should be kept ahead of concentrates. During this transition period the horse may need less corn than it was getting in the field.

Immediate Method

This is the old method of total change from grass to stable and is made in a day. Traditionally, the horse was then purged with an aloe ball applied with a balling gun or tube. Some books still speak of dosing the horse with 'physic'. Today, such measures are rarely used. However, a diet of mashes and maximum precautions against colic are very necessary if the immediate method is adopted. Concentrates should be introduced and increased gradually.

Early Conditioning

Condition must be produced slowly, but it is important that the horse should not stand in during these early days. A horse allowed to stand in will get 'above itself' and its legs may fill. Even on rest days the horse should have an hour in the field, clothed in a New Zealand rug. Alternatively, the horse can be walked out to have some grass and a little exercise in hand.

This is a good time to have the horse treated for worms and bots. Such treatment should not be carried out on the same day as an inoculation is given. A second worm treatment should be given four weeks after the first.

The horse in the field has a thick coat, natural grease and subcutaneous fat to keep warm. A stabled horse needs clothing so as to use food to best advantage.

Hardening the saddle and girth areas is an important matter. The keys to success are: 'no sweat for a fortnight and everything very clean'. A numnah helps. The horse is fat after coming up from grass and the saddle will tend to roll. When exercising, extra care must therefore be taken to sit square, straight and still.

Rubbing the skin twice daily with surgical spirit will help to harden the saddle and girth areas.

There are several ways of exercising a saddled horse quietly without a rider. A mechanical horse-walker, a loose school, lungeing, long-reining or leading from another horse are all possibilities.

Early lungeing must always be done in large circles and preferably in a suitable confined space with good going underfoot. In this way the horse is subjected to the minimum physical stress. Pulling on the lunge rein or turning in tight circles may cause undesirable strains on the tendons and ligaments of a horse in soft condition. Condition cannot be rushed.

In practice, for most people with limited facilities, riding the horse gives the greatest control. For the first two weeks after coming up from grass, exercise should only be at walking pace.

Another advantage of riding from an early date, as has long been traditional, is that when got up from grass the horse is relatively weak. After a few weeks it is fitter and stronger, and this may cause problems for the rider when the horse is first backed.

Immediately after coming up from grass, the horse will only need half an hour's exercise or less a day at the walk. This will be increased gradually over the first fortnight. Slow trotting can begin in the third week. Gradually the work will extend to an hour or more. In the fourth week, cantering can be introduced, with some uphill trotting. Canters may be up hill by the fifth week and in week six they may be a bit quicker.

In the early stages of conditioning, the horse may knock himself or stumble. It is therefore wise to wear brushing boots in front and to consider knee pads for road work.

The secret of conditioning is to increase work gradually – and the eye and ear of the owner are the best guide. Every horse is different.

Roughing-off

At the end of the working season, the stabled horse will need 'roughing-off' so as to give body and mind a rest after an exhausting working period. Like conditioning, 'roughing-off' should be a gradual process.

The benefits take some time to accrue, and it is not beneficial to 'rough-off' a horse for merely a month or so. The changes induced by 'roughing-off' will, in these circumstances, merely upset and set the horse back. A short rest is best accomplished by going on to light, easy work, with plenty of time at grass. Horses in steady routine work do best with no days off.

Hunters will be 'roughed-off' at the end of the hunting season in March. During the summer there will be good grass to make them fat. For other horses, the process may be carried out at times when the grass is poor, and then extra corn must be given and, if the paddock is exposed, there must be shelter at night. The late and famous David Brock once summarised the 'roughing-off' of hunters in this way:

> Their rugs should be removed, one at a time, commencing about a week before they are to go out. Their corn should be reduced gradually, and finally knocked off completely. Their shoes should be removed. They should be cleaned rather than groomed.

This quotation summarises the essence of the procedure. 'Roughing-off' is based on four main reductions: clothing, grooming, hard food and exercise. All reductions must be gradual – a little less each day. It is ideal if the horse can spend longer in the field each day, initially with a New Zealand rug. Later, most horses can manage without rugs in a sheltered field. Once the transition is made, it may, depending on the time of year, be appropriate to give the horse plenty of hard feed while it is at grass. This will not apply to hunters!

Today, the horse's hind shoes are normally removed; good feet will certainly benefit from this. However, there can be no general rule: a horse with poor feet may remain shod, with regular visits from the farrier during the resting period. A horse with cracked front feet, for example, may need a pair of light shoes to prevent damage.

Where it is impossible gradually to increase the number of hours the horse spends at grass each day, the animal should be turned out early in the day and should be observed at intervals. The first turn-out is best done with a very hungry horse, which will at once settle down to eat. Over-exuberance is always a danger.

The roughing-off period is an opportune time to attend to splints or other bony enlargements, as well as to sprains and strains.

Blistering at this time is common. Firing is, however, under increasing scrutiny. In fact a recent investigation for the Royal College of Veterinary Surgeons concluded that firing should be discontinued. The question is whether the operation or the period of rest is the main benefactor.

12 The horse under stress

Because of the greater demands made for high performance on the competition horse, hunter, or point-to-pointer, the standards of management must be of the highest. The fundamentals of horse management are the same whatever the use of the horse, but in the case of the fit horse in work, additional skills come into play. In particular, an exceptionally high standard of fitness has to be achieved if the horse is to realise its full potential. Consideration must also be given to the stresses and strains, both physical and mental, to which the horse is being subjected.

Attention to the Basics

Stable

Each horse must be treated as an individual. Within the selection of stables or boxes available there will be a best box for each horse. For example, a horse that likes to know what is going on in the yard will be fretful if stabled round at the back. A restless horse will do better in larger quarters.

Bedding

Whatever bedding is selected, the emphasis must be on a dust-free environment. If using shavings or straw, it is advantageous to the horse if it is not in the confines of the stable when new bedding is shaken up. Indeed, when sweeping a concrete yard, some damp sawdust will help to prevent clouds of dust. Although cases of allergy to bedding are rare, a change of bedding may improve performance; for example, many horses will eat straw, thus giving them a rounder barrel which is not conducive to fast work.

Feeding

High-quality fodder is essential: it must be as free as possible from dust and spores. Unless the horse is predisposed to being gross, it is customary to allow hay ad lib. during the first eight weeks or so of training. As the galloping requirement increases and hard feed goes up, the quantity of hay allowed must be reduced.

The exact amount of hard feed fed daily must be based on the characteristics of the individual horse. Temperament, the horse's feed-conversion rate, whether the animal is naturally lean or fat, and whether it is greedy or a shy feeder, are all factors to be taken into consideration. Today, there are so many feeds to choose from that it should be relatively easy to find a combination which is suitable and palatable for a particular horse. Stocking a range of feeds allows one to cater for the individual. Additives should be chosen with discretion, bearing in mind the individual horse's requirements.

If food is to be used to best advantage, two points must be borne in mind. First, the aim should be to keep the horse free from worms and other parasites by having a good programme of routine medication. Secondly, teeth should be inspected twice yearly, as roughness may affect mastication; it may also affect the horse's placid acceptance of the bit.

Feet

A reliable and kindly farrier is essential. To consult and discuss the horse's progress and achievements with the farrier may well increase the latter's interest. Observation by the rider on how the horse moves during fittening work may indicate that shoeing changes are needed. However, one should never dictate to a farrier; he must feel that most suggestions are his own! Consideration should be given as to when the horse will first require stud holes and the farrier should be advised in ample time.

Back

This is a controversial subject. Whereas some maintain that the horse's back cannot usefully be manipulated, others offer their services for that purpose. When a horse has gait imperfections that are detrimental to its performance, it is up to the individual horsemaster to decide whether to seek this line of help.

Legs

The regular use of stable bandages is not generally beneficial except in the case of older horses or where there is some other particular reason such as windgalls. This is because the horse-master would see the horse's legs only in their post-exercise or bandaged state. Where bandages are used, the modern thermal bandages are highly effective.

An astringent paste can be used after stress or to reduce minor swellings. It is best not used if there are any open wounds. Similar care must be taken if using astringent spirit or legwash. Bandaging the horse for the homeward journey is beneficial after particularly hard work, especially on hard ground. Such work is tiring for the legs, and tired legs are more likely to become damaged.

The great fear of all horsemasters is that the horse might develop 'a leg'. This may vary from a slight strain to a breakdown of a flexor tendon. Whatever the cause – be it a jar, a blow, or stress – the initial treatment is the same. The reaction to any suspicious heat or swelling is important. If in doubt, action should be taken immediately to prevent the problem getting worse. It is essential to stop work entirely and treat the problem until all heat and swelling are gone. This may take days or even months. If the condition persists, the vet should be called. While the horse is not working, feed must be drastically reduced and the horse should be led out in hand each day to graze.

If the leg returns to normal in three or four days, walking exercise can be recommenced. The horse should be trotted in hand to check for soundness. If the horse stays sound, work can be gradually built up again over the ensuing week. During the period of damage and recovery, both the leg and its neighbour will have been bandaged in the stable. The bandages may now be left off but the leg should be watched carefully for any sign of recurrence. Such setbacks will inevitably extend the horse's fittening programme.

Fittening Work

During the early part of the fittening programme, a particularly close eye should be kept on the horse's legs, saddle and girth areas for any signs of wear or soreness. If the legs start to fill, this may be due to too much heating or feed or to too fast an escalation of the work programme. It is advisable to start work with a numnah, even if later this is omitted. The numnah must be kept particularly clean. Numnahs with rubber fillings should be avoided as they do not allow the back to breathe.

Lampwick girths are soft and not so inclined to make the horse sweat or sore. If the horse usually wears boots, it is best to avoid rubber-lined ones as they may cause irritation. A quarter sheet should be worn during cold weather for slow work if the horse is clipped out.

Although in theory the horse should be kept at the walk for the first two weeks of the programme, if the animal misbehaves, either the feed or the work must be adjusted as there is no point in horse or rider getting hurt before the Season starts. Most horses benefit from time spent in the paddock each day as it helps to keep them on an even keel. After two weeks, the horse can begin some easy trot work. Traditionally, work in the manège or school is reserved for competition horses. However, improved suppleness and balance must be beneficial to all.

Such schooling should be at the rising trot, in not less than 20 m circles and with easy changes of rein. The horse must work in the correct outline for the proposed competition stage, but the exercises should be kept simple. Even during the preliminary roadwork the horse will have been asked to accept the bit so that all improvements are gradual. At this stage lungeing may usefully be introduced but it should be limited to 20 minutes daily.

During this period the horse should not be losing weight too quickly and diet must be watched accordingly. Suitable feeds may exclude oats but might include rolled barley, horse and pony cubes, flaked maize, a little bran and sugar beet pulp, and a boiled feed of linseed and barley perhaps twice a week.

The horse may need to be clipped, but this will depend on the thickness of its coat, the warmth of the stable and the animal's tendency to sweat. The choice of clip will depend on personal preference and individual circumstances. The use of rugs must be adjusted accordingly. Many people like to leave the legs unclipped because the leg hair affords some protection against such evils as cracked heels and mud fever.

During very cold weather, the horse will appreciate a deeper bed which comes right up to the door, thus stopping draughts. If paper or shavings are the bedding material, a low board may be placed across the inside of the doorway to keep the bedding in place.

In the fifth and sixth weeks, the horse can be doing up to 1½ hours on its roadwork days. More than this should not be necessary. Trot up any available hills but never down them, as this would cause unnecessary wear on the joints of the front legs. Those lucky enough to have tracks or grass on which to ride will cause less wear and tear on the horse's limbs and feet. Walking along the seashore through the waves is also very beneficial.

Cantering work should be commenced in the fourth week, starting with 1¼ km (¾ mile) slow canters building up to around 3½ km (2 miles) at the end of the eighth week. This should be sufficient for the average hunter or one-day-event horse. Specialised work such as dressage or jumping should be built up in weeks four to eight. This should preferably be at a level a little beyond that which the horse will be asked to achieve in competition. It is no bad thing for the hunter to receive some jumping training because the muscles for jumping will not have been used in that way for some considerable time.

A horse being prepared for a high-stress sport may usefully have a blood test occasionally so that the norms would be available for reference if trouble were to ensue.

The horse's vaccination programme is important. It should be phased to allow recovery time after each vaccination and is thus best done before the fittening programme is commenced. If this is not possible, or if such an opportunity has been missed, the horse should do no hard work for a week after the flu vaccination. The protection achieved by vaccination may mask the signs of infection, and great care should be taken not to damage the horse by working it when it is being stressed by disease.

Untypical coughs and sneezes in a horse should be taken as warning signs. Temperature should be monitored and work and feed cut back, and the horse should be kept warm. After 48 hours, if the symptoms persist, the vet should be called. It is essential to discover whether the cause of the symptoms is a cold, due to catching a chill, or a more virulent virus, a glandular infection or an allergic reaction. These all require different treatments and recovery times.

Performance Days

The Night Before

The day before hunting or going to a competition, the tack should be checked over, cleaned and put ready. If making a very early start next day, it may be a good idea to plait up.

When competing, the equipment should all be set ready the night before, and it is a good idea to have a checklist for both horse and rider. This should include all relevant rule books and paperwork. A timetable may usefully be prepared, working back from proposed departure time to starting morning stables.

If competing, the horse's performance at the competition will depend partly on how the animal spent the previous day, which must be carefully planned according to temperament and physical fitness. A highly strung or frisky horse may need to have an extra long work-out or perhaps a long session on the lunge. The horse should then be turned out to relax even though this may mean extra grooming. The resultant performance will make it worth while. The thick-winded horse may require a sharp 1 km gallop (around half a mile) the day before competing, although some point-to-pointers do this early on the morning of the race.

Final jumping practice is best two days before the competition. This gives the horse time to assimilate the lesson and allows treatment time for any knocks.

On the Morning

The horse should be fed early and at least one hour before leaving the yard. Water must be made available to the horse right up until departure time.

The horse should be groomed, plaited and dressed for travelling after completing its feed. If screw-in studs are used, the holes should be checked. Any road studs should be easily removable. It is very frustrating to get to a competition and be struggling to make the studs fit!

It is vital to check whether the horse has sustained any overnight injury or whether any other problem has developed. If at all suspicious, the horse should be trotted up to ensure soundness.

Ample journey time should be allowed so that one arrives at the meet or competition in a calm state.

Travelling Long Distances

Long-distance travel with horses is sometimes unavoidable. In practice, even on motorways, the maximum distance travelled should be no more than 350 miles a day. Long journeys should be broken by rest periods when the horses are fed and watered in the lorry. Overcrowding in the lorry should be avoided.

Long journeys can cause metabolic upsets and so the horse must be allowed recovery time before competing. During any journey, there is always a danger of a horse going down, and damage will ensue if the horse is tied up short. Any tie should therefore have a link of cord, which will snap in an emergency.

Horses are less likely to fret in strange travelling conditions if they are suitably distracted. They will settle better if they have had nothing to eat for the previous few hours and are then given a net of hay in the lorry.

On Returning Home

Once the horse is unboxed, it should be placed in its stable and have rugs and bandages removed so that it may roll if it wishes. The horse should have access to plenty of clean water and hay. It should be racked up and given a quick clean, especially around the ears, girth and saddle areas and between the hind legs. When cleaning the legs, it is important to check for wounds, heat or swelling, and to treat as necessary. The feet should be picked out, checking for over-reaches or stones wedged against a shoe. If all is well, the legs should be rebandaged. An anti-sweat rug should be put under the stable rug if the horse is unduly warm or is prone to break-out. The horse should then be given a small, easily digestible feed and be allowed to rest. Meanwhile, other necessary jobs such as tack cleaning and so on can be done, if the hour permits. A final check should be made before retiring.

The Next Day

Next morning, after feeding and mucking out, the horse should be unbandaged and should have its legs checked again. It is wise to have the horse trotted up to ensure that all is well. Strains, bruises and so on should be treated as appropriate.

Injuries are sometimes caused by badly applied bandages, which can impede circulation, as well as by boots rubbing against the flesh.

The horse should be groomed thoroughly, washing both the feet and the tail if necessary. Then, if conditions permit, a spell in the field is often beneficial; it is both relaxing and stimulating for the horse.

Some horses go off their feed after a competition or after hunting. They then need to be tempted back to the normal diet by being given several small feeds with added succulents – carrots, apples and so on. It does no harm to pander to the horse a little.

A number of horses become 'tucked-up' or 'run-up' after strenuous work. To keep such horses well and fit it may be necessary to restrict the number of outings, especially early in the Season. In any case, after a specially strenuous day, great care is needed. The horse should be kept on light walking exercise until all is well, and the key is always to build up work gradually again.

Stress

The horse encounters many forms of stress. Some physical stress is good because it allows the animal to develop strength and endurance. Other stresses can lead to distress. Anything causing stress is called a stressor, and the pressure may arise from a situation or a physical cause. Stress is the reaction to the stressor. It may be a physical breakdown, e.g. a stress fracture, or it may show as a syndrome of adaptive responses. These responses may be physiological (such as the release of adrenalin) or they may be psychological or behavioural.

The horse's responses are designed to cope with the problem and put an end to it. The trouble in stress is serious if the stressor continues or recurs so that the responses continue. The body is not designed to cope with such never-ending responses.

There are four main groups of stressors: Equine psychology, human psychology, physical causes and situation causes.

Equine Psychology

Physical conflict is common in the wild, but mental conflict creates greater stress. Common conflicts – for example, a horse that is frightened by an obstacle but is coerced by its rider – can give rise to later problems if they are mishandled. Uncertainty also creates mental pressures and confuses the horse, which then shows signs of its dilemma. It may stale or defaecate and such habits become repetitive.

Boredom is another stressor and is manifest by chewed wooden stables, fence rails and so on. It is possible that many vices such as wind-sucking and weaving are initiated by boredom.

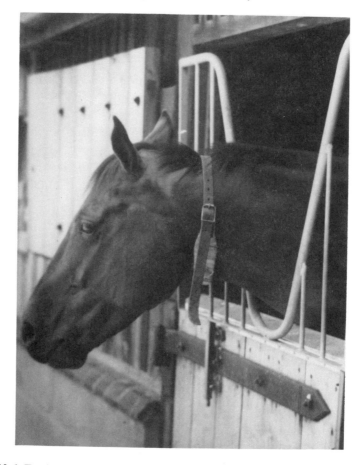

Fig. 12.1 Equine mental stress may be detected by 'bad habits' but it may exist and go undetected or be called 'sour', 'nappy' or 'a bad doer'.

The horse's temperament is an important factor. Temperament is probably governed most strongly by heredity, but is also affected by physical condition and mental development. In the short term, feeding is probably the most significant factor, e.g. a horse which is 'oated-up'.

Instinct is significant and must be taken into account. It is instinctive for the horse to get a predator off its back. In a moment of crisis it may do the same to its rider! The horse is gregarious by nature and can find loneliness a stressor. The right companions help the horse to relax.

Human Psychology

The psychology of human beings is an important stressor to the horse. This may range from 'killing with kindness' to sheer neglect. Fear is a dilemma for any rider, but presents a far worse dilemma for the horse. The bonds of trust are broken. A rider who is out of control may be terrified and 'saw' at the reins, causing pain to the horse. The horse forgets its training and reacts to the pain: a bolting horse is certainly reacting to stress.

Physical Causes

Many people thinking of stress consider only the physical aspect, and sometimes limit their horizons to stress that leads to fracture or other physical damage. There are many other physical causes of stress. Thus, speed, duration and weight of rider are all factors leading to exhaustion and fatigue, and it is the tired horse which fails to react quickly enough to protect itself at every footfall.

Environment is also important. Each horse is an individual in all respects, and environmental stresses vary with breed, age and condition. But there are certain basic rules. In light, airy stables with a good yard routine, the horse will lie down and relax during the day to a greater extent than in poky, dull and unpredictable yards.

Pain and ill health are clearly aspects where horses are at a disadvantage when compared with human beings. Diagnosis may locate the source, but it may be that many pains are unidentified.

Situation Causes

Birth is obviously the first stressful experience for any animal, and it is probable that early experiences affect subsequent events.

There are many artificial situation causes. Breaking, selling, or a change of home are obvious stressors. Hunting and competing – where the work is often hard – is very stressful and usually manifests itself in physical causes. Conversely, some horses are never happier than when working at high level. The key is in understanding the particular horse.

Physiotherapy

Today's horse must be fit enough to perform considerable feats and is thus exposed to the risk of injury. Physiotherapy is widely used during the training of human athletes, both to prevent injury and to aid recovery after it. These techniques are now being successfully adapted for use with 'the equine athlete' such as the hunter or competition horse. Obviously, it is more complicated to diagnose the exact site and type of pain in a horse and to gain its co-operation during treatment.

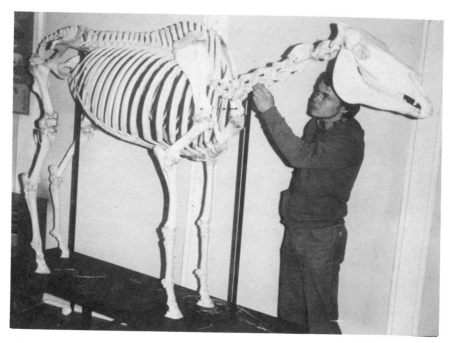

Fig. 12.2 A proper education concerning the horse is essential before attempting physiotherapy.

Traditionally, rest has been the treatment for the many athletic problems associated with horses. Now there are a number of techniques which, when used by competent people under veterinary supervision, lead to faster recovery, often of a more permanent nature.

A number of sophisticated medical appliances have been adapted to treat various equine athletic conditions. Some of these, such as faradism and ultrasound, obtain their effect by applying energy sources directly to the horse.

Muscles give stability as well as function to joints. Deterioration of joint function can be the result of a previous slight injury which may not have been noticed. The horse cannot control or operate the joint as well as before, but is not lame. If not noticed, the next time the joint is stressed it may malfunction and sustain a serious injury. Permanent damage may perhaps result, or the horse may show poor form due to discomfort. This could be the start of a nappy horse.

Faradism

Faradism can be used for both diagnostic and therapeutic work. It consists of the application of an intermittent alternating electrical current, allowing the muscle in question to contract and relax, thus preventing atrophy. The artificial stimulation applied by faradic current induces circulation to the injured parts without any damaging effect, so giving an ample supply of blood carrying nutrients and oxygen just when and where they are most needed. Faradism can reduce adhesions and promote the interchange of fluids within the body with beneficial results. It can also be of great assistance in correcting muscle imbalance and releasing long-term muscle tension.

Ultrasonic Therapy

This consists of ultra-high-frequency sound waves above the normal range of hearing. These waves are produced by conversion of high-frequency electrical energy waves to sound waves by the crystal in the head of the instrument.

Ultrasound can penetrate to a depth of between 5 and 10 cm (2–4 in) and affects the tissue in different ways. The resistance met in the tissues by the sound waves induces heat, a mechanical vibration is produced and the waves stimulate a chemical reaction. All three assist in the dissipation of swelling and inflammation, the removal of harmful breakdown products, the reduction of swollen tendons, and the breakdown of adhesions.

Vibration and Massage

The therapeutic value of massage and vibration has been known for many years. Their greatest value is in producing complete relaxation and dispersing muscle tension. Muscles cannot develop if they suffer from constant tension. Horses are generally exposed to unnatural ways of living and working which can induce muscle tension. The temperament of the individual horse is of importance in its attitude to its work and to its owner/trainer. Muscle tension can prevent the horse from relaxing mentally and physically as muscles held in undue tension become exhausted. In this state they cannot develop and this may lead to muscle imbalance or incorrect development in shape or form. The elimination of undesirable muscle tension as soon as it is observed enables the trainer to progress with the horse's training programme and helps to preserve the animal's co-operation.

Remedial Exercises

Injury can reduce the range of movement in joints, and by reduction of ligament suppleness or painful adhesions it can cause limitations.

Specific exercises can build up joint suppleness, help to break down adhesions slowly and to restore the horse's confidence in the use of the joint, and remedy muscle imbalance resulting from previous injury or use.

Heat Treatment

Heat in therapy can be used in three forms, radiant, conductive and conversive. The physiological effects of the three methods are basically the same and range from superficial to deep. Radiant heat is applied by means of infrared light. Conductive heat is applied by hot-water bottles, electric heating pads, hot fomentations and poultices. Conversive heat is developed in the tissues by resistance to high-frequency electrical energy.

Heat promotes circulation in the area, which aids healing and also helps to alleviate pain; thus it relaxes and soothes the patient.

Hydrotherapy

Generally the application of water is a means of applying cold. However, equine swimming pools allow the heart and lungs to be kept fit without any weight on the legs.

Nursing

Care and attention to detail has a psychological as well as a physical benefit. As with people, horses respond in a happy relaxed atmosphere with a positive attitude, gain confidence and become willing patients.

Part IV

Stud Management

13 The stallion

The same basic rules of stable management apply to stallions as to other horses. Stallions are often big and strong, and highly motivated to achieve their sexual objectives, and are sometimes temperamental. Like other horses, they benefit from a contented mental state. The route to safety lies in well-thought-out routines, and with calm, confident but firm handling and an insistence on discipline. The correct balance between permitted high spirits or enthusiasm and stepping out of line needs fine judgement which comes with experience.

Fig. 13.1 Mutual esteem between stallion and handler is the ideal relationship.

General Care

Accommodation

Like all horses, stallions are gregarious and should be kept where they can see other horses and some activity. Many stallions are kept successfully in a mixed yard, although at most Thoroughbred studs the stallions have a separate range of boxes. The stable must be sound and strong, and extra care must be taken to ensure that there are no sharp projections or edges as an excited stallion may be quite frisky in his box. The bars at the windows must be sufficiently close together to prevent the stallion getting a foot between them.

The box should be of a generous size. Some Thoroughbred boxes are almost 6 m (20 ft) square. The lower door may be higher than average in order to discourage the stallion from getting his forelegs over it. Many stallions are permitted to look out of the top door except when other horses are being led past. If bars are necessary across the top door, it becomes more important that the horse should not feel caged in. In any case, a solid top door must be available. If standing at public stud, the stallion will get very excited each time a horse-box approaches as it may be bringing another mare for his attention.

Exercise

The stallion needs daily exercise. During a busy covering season, the amount of exercise may be reduced but the stallion should never be confined to his stable and brought out only to cover a mare. This would be bad both physically and psychologically.

Many stallions are ridden at exercise, but they obviously need a competent rider. The degree to which they are ridden in mixed company depends partly on the time of year and partly on the degree of discipline to which they have been trained. In some countries, entire horses are more common than in others, and so partly it is a matter of custom. Risks should be avoided, and even when leading in hand the leader should plan his route carefully, wear gloves and carry an in-hand whip. In addition to exercise, stallions may compete, and the extra fitness may enhance their stud performance. Competing will also help to advertise a stallion.

The most valuable stallions are Thoroughbreds with a good pedigree and track record. Although these horses are, in general, handled and cared for brilliantly, this type of management is not a pattern for all. The high value of such animals necessitates minimum risk management and, since the Thoroughbred is more hot-blooded than other breeds, he is less easy in general.

All horses appreciate being turned out in a paddock, preferably every day. The stallion paddock must have rails which will not be jumped. It can, with advantage, have rounded corners and it may need some sight screening so that neither the stallion nor passing mares should be unreasonably aroused. Time will be saved if the paddock adjoins the stallion box.

In addition to ordinary exercise, a stallion may usefully be lunged, long-reined or loose-schooled if the facility exists. In winter, many stallions are left in the field with the mares which are in foal to them. A gelding may get savaged if turned out with a stallion, unless they were brought up together. This danger is increased in the presence of mares.

Ideally, the stallion should have plenty of human contact. The stallion needs consistency of treatment with fair reward and punishment administered mostly by voice. A relationship of mutual esteem is ideal.

Diet

A popular stallion will lead a vigorous life during the season and so must start it fit and strong. The stallion's demands for vigour and semen production require a diet similar to that of the competition horse.

Fresh water must be available at all times. Grass or hay should be of good quality, and concentrates (which must be adequate in minerals and vitamins) should include about 10 to 14% protein.

Health

Even when the stallion is living in, routine worming is needed. It is important that parasites sapping the stallion's energy are reduced to a minimum. The stallion should also be injected against tetanus and equine influenza. Teeth need checking regularly, and the stallion's feet must be kept in good order. If the stallion is doing much roadwork, he will have to be shod.

Proper strapping will promote good condition as well as making the horse feel good and look more attractive to the owners of mares.

Types of Management at Stud

There are two basic methods of standing a stallion at stud. In the
first, the stallion is segregated and covers the mares in hand. The
second method is for the stallion to run with the mares at grass.
Where this method is used, many stud managers prefer to cover
new and especially maiden mares in hand for the first time.

The stud season is from 15 February to 15 July for Thoroughbreds and a little later for other breeds. Outside the season, a
stallion may run with barren mares. When all else has failed, this
may get them in foal. However, with some stallions it is unwise to
turn mares out with them until the mare is in season.

Mares with foals at foot are normally covered in hand and
turned out as a group in a separate field. Because of the extra risk
involved, a Thoroughbred stallion would rarely be turned out with
mares. However, for most stallions it is good that they should learn
to live in the company of others, as is the case in the wild.

A stallion running with mares will, as a youngster, get well
kicked. In this way the stallion learns to be more careful and
discerning, to approach the mare from the side and to enquire
before he mounts her. A stallion running with experienced mares
can learn his job and will tend to be more sensible when covering
in hand.

Use of Stallions

Until recently in this country, most stallions were licensed by the
Ministry of Agriculture. It is now left to the Breed Societies to
make their own arrangements for validating the use of a horse at
stud. Breed Societies may require an inspection, which may
include conformation to type, freedom from defects, hereditary
disease and vice, and a check on fertility. If a two-year-old colt is
to be inspected properly, it is important that he has been taught
good manners and can be shown off in hand at the time of the
inspection.

Stud Fees

The amount charged as a stud fee is a matter for the stallion
owner's discretion. There are several standard arrangements which
may affect the ultimate fee. Care should be taken to make the
price competitive, especially if the stallion is not well known and
there are others of the breed in the area. On the other hand, to

undervalue the stallion may be taken as an indication that he is inferior. Producing a foal is expensive and it is a false economy to save on the stud fee. Mare owners should seek the best stallion that they can afford. As the mare will often stay at the stud for some time, the charges for keep are an equally important cost item.

If a stallion is not 'special', he should not be standing at stud. The stallion's breed, blood-line or sheer quality should make him of special value and so enable him to be advertised with confidence.

The most common arrangements for payment are:

(a) A straight covering fee of £x, normally payable before or when the mare is collected from the stud.

(b) 'No foal, no fee': if the mare is tested on an agreed date and found not to be in foal, the covering fee is returned to the owner.

(c) 'No foal, free return': the fee is payable as in the first case, but if the mare does not hold that year, she will be covered without further fee the next year.

(d) 'No live foal, no fee': if the foal does not live 48 hours, the fee is refunded.

(e) 'Split fee': part of the fee is paid for the mare to be covered, and the balance is payable under an arrangement such as (b) or (d).

The cost of the service offered to the owner of a mare tends to get higher as the risk decreases. For example, if the straight fee is £100, one might pay £120 for 'no foal, free return' or £150 or more for 'no foal, no fee'. Concessions may be offered to the owners of approved mares with a good breeding record. Mares which are difficult to get in foal tend to spoil the stallion's fertility figures. An owner will wish to claim that, say, 80% of the mares covered get in foal. The more selective the stallion owner can be about visiting mares, the better will be the stallion's performance figures. The owner of a new stallion with no proven record will need to offer some concession.

Thoroughbred stallions are often owned by a syndicate of owners. In a typical case, the stallion's value might be divided into 40 shares, with the owner of each share getting a free nomination, which may or may not be transferable.

Stud Facilities

A stud where the stallion runs with the mares will need no special facilities. However, on most studs there is a 'teasing' or 'trying board' where the stallion can tease or try the mares to see if they are in season. Although some owners will tease a mare at the stallion's door, there is then a danger that the mare will strike out and hurt herself on the door catches.

An alternative is a hatch where the stallion can put its head out and the mare stands, possibly in a crush or race which confines her. The wall below the hatch should be well padded, and similar protection should be affixed to the opposite wall towards the rear of the crush. An alternative arrangement is a free-standing padded wall (Fig. 13.2), well secured into the ground. This should be set on a free-draining and non-slippery ground surface. If protected from the worst of the wind and rain, this is an advantage. The area used should be free from disturbance and out of public view.

Fig. 13.2 A stallion at the trying, or teasing, board and a mare who is not in season.

In some studs, an indoor teasing board is used. This is often hinged out from the wall of a barn. Indoor facilities are a particular advantage for Thoroughbred studs which operate earlier in the spring because of the need to produce well-grown yearlings by the

January following the foal's birth. In a stud barn, the floor will be a 'tan' mixture such as is used in a riding school, and the surface must be kept rolled level and watered so that it is firm and dust-free.

The board itself should be very robust and of a height such that it is level with the dip in the back of the average mare. The board must be longer than the mare so that she cannot kick the stallion or strike out in front of him. Some studs have a small pen at the front, offset from the line of a teasing crush. The purpose of this pen is to enable a groom to hold the mare's foal safely near her head. Some mares may be distracted if their foal is left away from them.

Another feature of a crush may be a bar at the front. This stops the mare running forwards. There may be a second bar at the back for use when the mare is being prepared for service or is undergoing veterinary examination.

In addition to teasing mares in the stud yard, it may well be necessary not only to walk the stallion past the mare's field but also to try mares over the fence. For this purpose, a plain close-boarded teasing board is set in the fence. At one time, protection on the top of the board took the form of a large wooden roller in case the stallion got a foreleg over the top, but anything safe and smooth will suffice as protection, and heavy duty rubber is commonly used.

Use of a Stallion

Many colts are sufficiently mature at two years old to get mares in foal and it is all too easy to think of the two-year-old as just a youngster and fail to take proper precautions against the risk of the colt popping over the fence and covering a mare. The three-year-old may be permitted a few experienced mares. The four-year-old may be advertised and may take a dozen, or even up to 20 mares. A five-year-old stallion should be able to cope with twice that number. However, it must be remembered that Thoroughbreds, and horses which have been done well, mature faster and tall horses mature more slowly, so allowances must be made.

Promotion

The stallion must be advertised. Advertisements should appear in the local press, in special stallion numbers of weekly and monthly magazines and in Breed Society publications. The stallion may also be paraded at shows or else may compete where there are suitable classes. On such occasions the handler and any assistants should hand out printed stud cards to those interested, explaining the particular virtues of their charge.

Where there are several stallions at stud, an annual open day and stallion parade is a means of advertising.

Trying

'The Right Moment'

If mares are being covered in hand and the stallion has many mares, it is best that he should cover a mare as rarely as possible. Covering is time- and energy-consuming. The object is to cover the mare just before she ovulates so that the sperm going up meets the egg coming down. The difficulty is to know the precise moment. In the field the stallion repeatedly covers the in-season mare as long as she will receive him and he has the energy.

In a Thoroughbred stud the vet will put his hand in the rectum and feel the ovary through the gut wall. On some studs, the stud groom may observe the cervix through a sterile speculum.

On most studs, past experience of the particular mare will give a clue as to how long she will be in season. She will ovulate in the last two days of being in season, so, if the mare is going to be in season for five days, there would be little point in covering her before the third day. The semen of most stallions will still be viable in the mare 24 hours later and so to cover her again on the fifth day would be merely a safety measure.

However, even if the mare worked to such a typical pattern, the success of the plan depends on knowing exactly when the mare comes into season. This knowledge is essential and is gained by keeping accurate records and by observation. These two procedures are backed up by letting the mare know that the stallion is about. Although some well-mannered stallions are taken into the mare's field, the normal pattern is to walk the stallion in a bridle past the field where the mare is.

When the mare is in season she can be tried by the stallion. However, trying a mare will involve the stallion in some slight risk and so very expensive stallions will not be used. This job will be done for them by a rig or a low-value stallion called the teaser. Trying the mare is much easier with an experienced stallion and such a horse will sometimes show very clearly by his behaviour whether the mare is ready or not. If the stallion trying the mare is well mannered, life is less fraught for the handlers. Some stallions are over-enthusiastic and do not make the best teasers. The actual process of trying or teasing the mare may induce her to come into season.

Where the mares are to be tried at a trying board in the field fence-line, the stallion or teaser is taken to the board. He will probably call to his mares. The stallion handler notes which mares come to the stallion and how they react. An assistant will fend off protective or over-attentive mares. Some mares are shy about coming to the stallion and the assistant should put a head collar on these and bring them up to the board to be quietly teased.

Trying Procedure

Each yard has its own procedure but the following is typical. The mare is led up in a bridle and stood behind the trying board with the handler standing by her left shoulder in the conventional manner. The stallion is then led up to the trying board in a bridle and lead rein. The handler carries a whip or stick and should wear gloves for safety. Some yards also insist on the handler wearing a hat: any hat gives some protection but a hard hat with chin harness is clearly the safest.

The stallion may make first contact with the mare nose to nose and the assistant must be on guard in case the mare strikes out. If the signs look hopeful, the assistant turns the mare parallel to the board so that the stallion can first nuzzle her right shoulder and then work his way back to the vulva.

Reactions that indicate a mare is in season include interest in the stallion, clear discharge from the vulva, slight squatting as she micturates and 'winking'. The vulva may be a little more filled with blood and look smoother and longer than normal. Although a mare may be coming into season and show some of the symptoms, she may not be ready to receive the stallion and will show this by kicking out, squealing, tail-swishing and laying back her ears.

Some shy mares will resist initially but, with patience, they will come round to the idea that they are ready to be mounted by a stallion. The stallion or teaser must therefore be willing to remain both enthusiastic and patient as he tries the mares. An over-noisy or aggressive stallion could prove off-putting to some mares but, on the other hand, a placid slow stallion might not be sufficiently stimulating.

Covering

When a mare is covered in hand, there are two styles common at studs. The main difference between the styles is accounted for by the conflicting requirements of being natural and being safe. To be totally safe, the mare may have hobbles put on her hind legs and these are attached by very strong cord to a neck strap. Other safety precautions include blindfolds, holding up a front leg, and twitching the mare. At the other end of the scale, the two horses on long lead reins are allowed to meet and mate in their own time and own way.

Procedure

A reasonable compromise is as follows. The upper hindquarters of the mare are groomed so they are free from dust and dirt and then they are washed with mild disinfectant, paying special attention to cleaning the dock, vulva and anus. Several swabs are used so that a dirty swab is never put back into the clean liquid. The tail is bandaged with a freshly laundered tail bandage. The long hair may be plaited and turned back into the bandage. The important point is that a tail hair should not lie across the vulva as the stallion's penis is inserted.

Felt overboots are then strapped on to the mare's hind feet to lessen the risk of her kicking the stallion. She is then held in a bridle and a long twitch is left ready in case it is needed. The stallion in his covering bridle with its lead rein is brought into the covering yard (Fig. 13.3). He should be brought to the mare's left flank where he can nuzzle the mare quietly and gain her confidence. If the stallion were to approach from behind, he might get kicked.

As the mare gets well aroused, the stallion will be similarly aroused although each stallion has his own speed of operation. If the mare is inclined to kick, a twitch can be applied for extra

Fig. 13.3 A well grown three-year-old stallion showing good manners.

safety. In general, it is not necessary to twitch a mare, but with high-value stallions, twitching may be used as routine.

The smell of the mare causes the stallion to draw back his upper lip with his head held high in the Flehman posture. By the time the stallion has worked his way to the mare's rear, he should be fully drawn, i.e. he should have a fully erect penis. He may then mount. At this point some handlers like an extra assistant to hold the mare's tail out of the way and even to put the penis into the vulva. However, most stallions and mares can cope quite well.

As the penis goes up into the vagina, the mare may step forward a pace to balance herself. She must be kept straight. In the rare cases where the stallion is over vigorous with his front feet on her shoulders or his teeth on her withers, a protective leather may be put on the mare. A drop noseband is useful for stallions that are inclined to bite their mares.

Once into the mare the stallion will 'flag' his tail, which shows he is ejaculating. In his own time, he will quietly get down off the mare.

As soon as the penis is withdrawn, the stallion handler signals to the mare's attendant who pulls the mare's head and leads her left and forwards. This move automatically turns the mare's quarters away from both the stallion and his handler so that they do not get

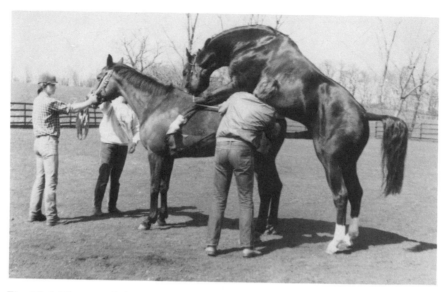

Fig. 13.4 The mare is kept straight and the stallion mounts her confidently.

kicked. The second assistant steps forward quickly with a jug of mild disinfectant to wash the stallion's penis. This must be done in the few seconds before the penis is taken back into the sheath.

As the stallion withdraws, a little semen may be lost from the mare. The kicking boots must be unbuckled at once and the mare is then walked on for some while until she is settled, because some mares wish to urinate after service and it is considered that the mare might thus expel further semen. The semen should be in the uterus but some may be at the top of the vagina by the open cervix. The stallion, too, may like to be walked round while he cools off.

Young, inexperienced stallions need quiet experienced mares of the right height, who will stand steady. If the young stallion mounts at the side, the mare must be walked and repositioned for him to try again. Patience and time may be needed. A very enthusiastic youngster may require a second line to help restrain him. The second handler on the other side may become at risk if the stallion or mare turn out of line.

Where there are significant height differences between the mare and the stallion, it may be possible to use a slope or bank to help the stallion or even to stand a big mare in a hollow such as the water jump.

The young stallion will cover a mare a day; the older stallion will cover a mare in the morning and another in the afternoon. However, even covering mares only on alternate days, it may be that several mares will need to be covered in a single day and then the stallion must cope with these demands. The stallion who was got really fit before the stud season begins, and who is kept fit and strong through the season, will be better able to cope and is more likely to get a higher percentage of his mares in foal.

Getting all the Mares in Foal

Keeping a mare is expensive. There will be a period of at least 18 months from the day the mare was swabbed prior to covering until the foal is weaned. In addition to the costs of keeping mare and foal, there will have been the stud fee and related charges, and possibly some veterinary expenses. Producing a foal is therefore an expensive business and during the 18-month period the mare's value may have decreased. If, instead of producing a foal each year, the mare produces only two foals in three years, most of the costs are increased by 50%. It is, therefore, very important to ensure that each mare going to a stallion should be got in foal.

The first step is to spot the mare coming into season so that she can be swabbed by the vet. It is also important to identify when the mare next comes into season so that she can be tried at the right time. The mare should normally be covered on the second day of the season and thereafter on every second day until she 'goes off'. Some mares only stay in season for a few days, and they must be tried and covered daily. Other mares, particularly older ones, appear to have rather acid conditions within the genital tract and the sperm seems to fail to survive; these mares also need serving daily.

The problem mare is one who comes into and stays in season but fails to ovulate. An injection of luteinising hormone will cause the egg to be released from the ovary. A mare that fails to come into season can be helped by having the genital tract irrigated with saline solution. This should be done by the vet. Some mares will need a prostaglandin injection if a blood sample shows progesterone in the blood. Others will need progestogen, which can be given in the food for 10 days. Progestogen acts like progesterone, and when the treatment is stopped the mare will come into season some five days later. Usually she will ovulate on the fourth day of her season.

Once a mare has been covered and has gone out of season, it is important to ensure that she holds. She will therefore be checked to see if she comes into season again. A mare that has been properly covered but has failed to hold will need veterinary palpation of the ovaries to ensure that they are in a normal condition.

Forty days (six weeks) from the last service, the vet can test the mare by manual examination to see if she is in foal. It would be unwise to do this at any earlier stage because the foetus would not be properly attached. Alternatively, after 10 weeks, the vet can take a blood sample so that pregnancy can be diagnosed. A third means of testing for pregnancy is by means of a urine sample taken after 17 weeks. The newest method for detecting pregnancy is by special electronic ultrasonic equipment.

Paperwork

Memory is a fallible thing and consequently proper records should be kept in respect of each mare. These should be an ongoing record of the mare's performance at the stud, with particular reference to the days the mare was tried, in season, served, and tested in foal (or not), and when she foaled. Other items noted will be farriery and veterinary treatment.

14 The mare

Selection for Breeding

There is universal agreement that the wrong reason to select a mare for breeding is because she is not fit for anything else. A good reason to select a mare is because she is an outstanding example of quality. It may be that she has proved this quality in competitions, be it in the show ring, the racetrack or elsewhere. Usually it would be wrong to go on breeding from a mare that has difficulty in holding her foetus or in foaling, or from one that is a bad mother. Only an exceptional mare would justify these extra problems.

It is also wrong to breed from a mare with poor conformation or a poor temperament. Certainly, sentiment is no justification for breeding from such a mare. It would be wrong to breed from a mare if her foals are not full of quality, and are of good

Fig. 14.1 Good brood mares. On the left a warmblood and on the right a Thoroughbred.

conformation and temperament, and robust and healthy. All too often a mare is bred from because she is there and not doing anything else. There are too many horses around and there is only room for the best.

Not all brood mares are beautiful. A mare may be a good type, but lack quality: when crossed with a suitable Thoroughbred that mare will produce superb offspring. In Britain, there is discussion concerning a National Riding Horse because, in spite of originating the Thoroughbred and having superb native breeds, when it comes to sending out National Teams to compete in driving, dressage, showjumping and eventing it is mostly imported horses that are used. The ideal is a home-bred horse with the courage of a Thoroughbred, the good temperament and strength of an Irish Draught, and a combination of quality and presence. For showjumping and driving the stronger middleweight version would be used and for eventing the faster lightweight version; for dressage one would look for movement and good temperament.

The traditional English hunter is such a variable commodity that it is hard to standardise the type, even though, through the work of the National Light Horse Breeding Society (formerly the Hunter Improvement Society), most hunters are now sired by Thorough-bred stallions.

Sadly, the qualities now being sought are those that have been frittered away through lack of foresight. The Yorkshire Coach-horse was amalgamated with the Cleveland Bay, the Devon Packhorse disappeared, and so few Norfolk Roadsters survived the war that they could only point to what might have been. Even now no national effort has been made to save these blood-lines, which soon will have gone for ever. The only draught-lines remaining, other than the cobs and carthorses, are the Hackney and the Cleveland Bay. The former has been bred for action, and the latter, although it has the size and bone required, seems to lack the zest and willingness of spirit to meet with general approval for competition work. There is thus a great shortage of British stallions to cover Thoroughbred mares for crossbreeding.

Where pure breeding is proposed, the task is made easier by Breeds, other than the Thoroughbred, having showing classes where skilled judges can indicate their personal preferences.

Selecting a Suitable Stallion

There are three aspects to consider when selecting stallions to look at for a particular mare. These are genetic potential, performance and progeny.

Genetic Potential

Genetically, the stallion is half of each of his parents, one quarter of each of his grandparents and one eighth of his great-grandparents. The information concerning his relations is generally available and is worth considering because his offspring will be half of him genetically. Within the Thoroughbred 'General Stud Book' there were 50 mares with traceable offspring in the first issue and this gives the 50 numbered families. In any breed, to breed within a line or family is called inbreeding and this will intensify both good and bad points. Inbreeding can be quite safe where the genes of the family are known to be pure, e.g. the Exmoor pony. Inbreeding can also bring a recessive and undesirable characteristic to the surface. Consideration of any pure siblings (brothers and sisters) of the stallion will show the same genetic potential in a different light. Consideration of half-siblings is also relevant, as a study of a Thoroughbred sale catalogue will show.

Performance

What has that stallion done? Consideration should be given to his competitive achievements, his winnings and whether he stood up to training. If the stallion was a racehorse, what sort of distance was he best over? Has he a good reputation for temperament? Does he move well? Is he sound? Has he good fertility? These are all relevant questions.

Progeny

The stallion's progeny must also be considered to see whether he throws desirable characteristics and stamps his type. It may be that the offspring have undesirable traits or conformation faults.

Having shortlisted two or three stallions which are suitable for the mare in question, they then have to be visited. At the stud, the conformation, movement and temperament of the stallion should be studied. Another very important consideration is the standard and style of the stud. The mare owner must feel confident that the mare will be well cared for and pleasantly treated.

Condition of the Mare

When a mare goes to stud she must be in good condition. Her skin must be clean and free from parasites. Her feet must be in good order, recently trimmed and unshod behind. If she has not been very recently wormed, then the stud may worm her on arrival. Many studs like the mare to be vaccinated against tetanus and equine influenza and they may require documentary proof.

In terms of body weight and condition, it is difficult to breed from fat mares or very thin ones. The mare should go to stud fit and well, with a reasonable cover over her ribs and on a rising plane of nutrition. This slight increase in feeding mirrors the natural increase in spring grass which is part of the trigger mechanism to start the mare cycling regularly. Improved weather giving her sun on her back is a second trigger factor, but the most significant is the longer day.

With Thoroughbreds it is desirable to have early foals so it is necessary to start the mares cycling in February. To trigger the system, from January onwards bright lights are switched on in the mare's quarters to give at least 16 hours of light per day.

When the mare arrives at the stud she has to adjust to different feeds, different water, and even different germs. Thus, when she first arrives she will regress slightly in condition, and it is therefore wise for the mare to arrive at the stud a week before she is next due in season. This aids the chance of getting her in foal. However, if it is required to cover her at the foal-heat, it is not wise to travel the mare and newborn foal so it is best for her to foal at the stud. In order to ensure that a mare is not too heavy, and to enable her to acquire immunity to the germs in her new environment (and thus pass on this immunity to the foal), she may go to stud as much as a month before foaling.

Artificial Insemination (A.I.)

With the aid of a suitable mare it is not difficult to collect semen from a stallion by means of an artificial vagina. The semen can be stored at a very low temperature. It may be diluted, or it may be used in its natural state.

There are several benefits of artificial insemination, notably disease control. If the stallion does not physically touch the mare he cannot contract any form of venereal disease. The contagious equine metritis epidemic in 1977 had a profound effect on the finances and reputation of many Newmarket studs; hence the stringent and expensive precautions now in force on most studs.

The second major benefit of artificial insemination is reduced risk of injury to the stallion. He never mounts a strange mare and this avoids his being kicked in the penis or testicles and so of being put out of use.

The third major benefit is that the semen is actually placed up through the cervix into the uterus of a mare at the right time.

Because of these benefits, many people use A.I. to convey semen from their stallion to mares. The possibilities of A.I. have attendant disadvantages. Because semen can be diluted, one stallion could father thousands of foals in a season, thus making hundreds of stallions redundant and putting most studs out of business. The diluted semen is held in special straws. Moreover, to ensure that the mare was in foal to the right stallion, all ensuing foals would need to be blood-typed.

Legislation is probably needed to provide the necessary safeguards if artificial insemination is to be widely used. 'Test tube conception' and surrogate mothers are another future possibility.

Disease Control

Internal and external parasites must be controlled as routine. The mare should be vaccinated against tetanus and equine influenza a month before foaling. Diseases affecting the genital tracts may interfere with the reproductive processes and be transferred from mare to stallion or vice versa.

At one time, mares and stallions were assumed to be 'clean' in the absence of discharges or other evidence to the contrary. Today, the owners of studs like to be sure. In particular, they wish to be certain that the mare has not got contagious equine metritis (C.E.M.). This disease is caused by bacteria which may be found lurking around the clitoris.

When the mare is in season and thus has her cervix relaxed, the vet can take swabs from the genital tract and these can then be tested for C.E.M. and other diseases. Each stud will set out its own conditions for the season.

Care of the In-foal Mare

Management of the in-foal mare divides itself into three over-lapping periods. During the first two months it is essential that the mare should not have a fall, blow or experience that might disturb the embryo. Once all is secure, there is the middle period of the pregnancy when the mare must simply be kept fit and well. Finally there is the last three months before foaling day.

The Early Months of Pregnancy

Hopefully, when the mare does not come back in season after being covered, she is pregnant. The pregnancy can later be confirmed by the vet making a manual examination, by blood or urine samples, or by machine (see Fig. 14.2).

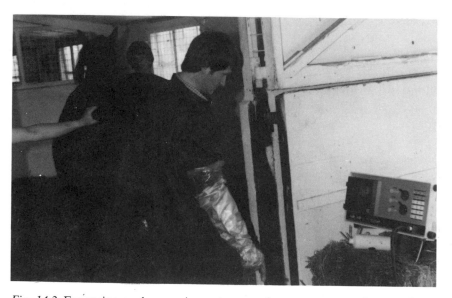

Fig. 14.2 Expensive modern equipment can confirm pregnancy at three weeks instead of waiting six weeks.

When a mare is not confirmed in foal, it may be that she did conceive but the embryo or foetus failed to take up a good anchorage, died and was reabsorbed. Germs infecting the mare's genital tract can be another cause of loss: there are various viruses, bacteria and fungi that may cause abortion. Unfortunately, an abortion at an early stage of pregnancy is not easy to detect. The small heap of reject material is quickly reduced by natural scavengers and will go unnoticed.

The conformation of the vulva of some mares is such that the lips are slack or incorrectly aligned. This defect can be remedied by Caslick's operation.

A further possible cause of abortion is that the mare has conceived twins. If this happens, in the majority of cases the mare will abort. If a mare carries twins to full term, they rarely survive.

Although the use of pregnant mares is now less common, there is no reason why the mare should not continue to do ordinary light work for the first two-thirds of her pregnancy. Throughout her pregnancy it is important that the mare be kept fit. She should therefore take daily exercise. Many mares will remain out at grass and thus exercise themselves by grazing. Thoroughbreds and horses on heavy clay soils that poach may have to come in at night. Where there is no suitable field for grazing, daily controlled exercise is better than wandering around a straw yard. The 'horse-walker' machines have a useful part to play here, and twenty minutes' daily brisk walk would be a minimum for stabled mares. Alternatively, the exercise may be ridden or in-hand.

The Middle Months of Pregnancy

The mare must be kept in good order. Feet, teeth, and worm control will all be of importance. Some wormers are not suitable for pregnant mares.

During this period, the mare will go into her winter routine and feeding will change accordingly. The aim is to keep the mare fit and healthy but not fat. A mare which is over-fed at this stage will lay down fat in her body which will hinder foaling and increase the strain on her legs. However, feed must be good. It must contain adequate minerals, vitamins and essential amino acids. It must be mould-free as some toxic foods will cause abortion.

As winter approaches, the mare's owner must ensure that the hay fed is clean, well made and well stored. The concentrates may be a cereal with a proprietary supplement which may include vitamins A, D, E and folic acid as well as minerals.

If the mare has a foal at foot, supplementary feed may be avoided until weaning is complete and the milk supply has dried up. At weaning, mares may be stabled for a day or two and fed only on hay or they may be turned away on to a bare, well-fenced pasture with a companion. The supply of milk should start to dry up after a few days. The udder should continue to be checked for mastitis.

The Last Three Months of Pregnancy

During this period, the main essentials are continued fitness, freedom from undue stress and a healthy diet. As the food is increased, it is best to change gradually to stud nuts. These contain the right level of protein as well as the essential amino acids. If a home mix concentrate is being used, the feed compounders produce a high protein balancer nut for adding to a home mix; it should not be used alone. Alternatively, grass meal and soya-bean meal or milk powder can all be included in the ration for the final month before foaling. This may be continued for a month or two after foaling, depending on the availability of grass. The stabled mare's diet should be kept reasonably laxative.

In the later stages of this period it is more important than ever to be familiar with the normal habits and behaviour of the mare. A change in the pattern may be the first sign of foaling.

Foaling

Foaling is a natural process which most mares would rather cope with unaided and undisturbed. However, things can go wrong and whether it is policy to foal indoors or out, a foaling box or mare-and-foal box may be needed.

Foaling Box

A foaling box must be larger than an ordinary box, and 4.6 m (15 ft) square is a good size. The reason for a big box is that when a mare goes down to foal she may lie sideways across the box, leaving inadequate room behind her. The box has to be cleaned out in advance and scrubbed clean. It should be disinfected throughout.

If a mare is going to be brought in when foaling commences, a thin layer of sawdust should be put under the bedding to prevent the mare from slipping. The box must be free of all sharp projections or obstructions. It should have a small night-light bulb for nights and be well lit for times of action: a dimmer switch is ideal. An infra-red heat bulb should be available.

An adjacent room where an attendant can sit up without

disturbing the mare is desirable. On large studs, closed-circuit television is used, but regular visits, although more disturbing for the mare, will provide a reasonable chance of viewing the activity.

A plastic beer-making bucket with a close-fitting lid will provide a suitable container for the requirements. A clean working smock should be available and it is good for the mare to get used to people wearing such a garment. Fawn is more practical than white as a colour. A place is needed for thorough hand-washing and nail-scrubbing. Hot water may be required.

The equipment ready for various eventualities at foaling might include the following:
Some clean towels to rub dry a foal if necessary
A foal rug (or old jersey)
An enema syringe, plus liquid paraffin
A torch (for foaling in the field or in case of power failure)
Some string
Water, soap and towel
A rug for the mare
Sterile foaling ropes in a sealed pack
Cotton wool substitute
Iodophore liquid (dairy teat dip) or antibiotic spray
Baby's feeding bottle and calf teat

Signs of Imminent Parturition

Mares carry their foals for about 330 to 340 days. The foal may therefore be expected eleven months from the last service, but an earlier or later delivery is not uncommon. The udder will usually have filled. A drop of 'wax' should form on the teats and milk may run. The muscles around the tailhead soften to a jelly-like consistency and sink. The quarters look slightly impoverished. The vulva normally looks firm and dry, and a hand placed across the hind quarters will not touch it. At this stage it will relax and moisten, and it may touch a hand placed across the quarters, showing that it has moved a little to the rear. Some of these signs can be seen in Fig. 14.3. Some mares will have people anxious weeks before the event; others may show only a slight change in behaviour, with increased restlessness.

Foaling

Foaling in the field is natural and is common for all native stock. A foaling box is common for Thoroughbreds. The foaling box offers

Fig. 14.3 Full udder, wax on teats, muscles round the tail-head soft, quarters looking less round, vulva relaxed and moist, and stood apart from the others – this mare did foal that night.

convenience. Checking a field at midnight with a torch is a slow business! A modern foaling alarm is a useful but expensive piece of equipment which gives peace of mind and saves many hours. It is attached to the mare and sounds a remote alarm when foaling starts.

Mares usually lie down to foal, but at any stage they may get up, walk around and lie down again. The mare will look round anxiously, even before there is anything to see. The mare with contractions pushes the placenta membrane out through the cervix.

The membrane then bursts, releasing the 'waters'. This first stage may be followed by a short rest.

The second stage occurs within ten minutes and is the appearance of the amnion surrounding the foal. This, too, will generally burst, releasing mucus. At this point many stud grooms like to enter the box quietly and check that the bulge showing at the vulva now contains a front foot, with a second front foot just behind it. Just behind and above the front feet there is the nose. This is the position for normal presentation and the attendant should leave the box quietly.

Even for this simple check, the attendant should be 'scrubbed up' or wear disposable surgical gloves. If the position is not normal or there is a delay in progress, then experienced help may be needed and the vet should be summoned. Most mares cope without help.

The front legs, head and then the shoulders emerge from the mare. The foal should be left to lie with its hind feet still in the vagina. This is important, as the mare will lie still and blood is being transferred through the umbilical cord from the mare to the foal. Possible action at this stage, if necessary, is to enter the box quietly and break the amnion covering the foal's nose, wiping it clear of mucus so that the foal can breathe. Normally, this help will not be needed as the foal breaks the membrane with its front feet. Although it is tempting to doubt the mare's and foal's ability to cope, nature generally knows best.

In due course, the mare will move, the foal's hind feet will be clear of the mother, the umbilical cord will break and seal itself, and the mare will recognise her foal. She will lick it clean and the mare–foal bond will be formed. A brief intrusion may be made to dip the naval stump of the foal into iodophore liquid, which can be obtained from the vet in advance (iodine is too strong). Dangling from the mare's vulva will be the umbilical cord and amnion. Sometimes the attendant will tie this up to itself so that it is clear of the ground.

A foaling box should not be warm at foaling time because the foal starts breathing as a result of its initial contact with the colder air. If a foal fails to start breathing, a bucket of cold water can usefully be thrown over its head and chest, followed by covering one of its nostrils and blowing up the other. When the foal is three hours old the stable may be kept a little warmer if mare and foal need to stay in.

Fig. 14.4 Leave well alone!

Two things must happen within six hours. The afterbirth (placenta) must be passed naturally by the mare to complete the act of foaling. If this does not happen, the vet should be called. If it does occur, it should be checked to make sure the horns are intact, and put in a bucket with some water over it so that if there are complications within the next 24 hours the vet can check to see if all the afterbirth has come away.

The second vital happening is that the foal must suck. Assistance is often given when it is not needed. If the foal does not get to its feet after an hour or so, some assistance may be given. However, just as a foal will fail to get up several times before it succeeds, so too it will fail to find the teat at first. Interference should be avoided if possible. If it is necessary to help the foal, an assistant may be needed to hold the mare.

The mare's first milk is the colostrum and contains the antibodies necessary for the foal to survive. It is high in essential nutrients. It is vital that the foal receive this and, as a last resort,

Fig. 14.5 Mare-foal bonding is important and best done without distraction.

the mare may have to be milked and the foal bottle-fed until it has the strength to cope on its own. If the mare was not given a tetanus booster 6 weeks before foaling, then the foal will need the vet to give it protection against tetanus.

The foal's first droppings, called the meconium, are delivered within the first 24 hours, provided the mare has been on a reasonably laxative diet prior to foaling. If the foal is unable to pass meconium, experienced attendants may try back-raking or using an enema syringe. Those with less experience will need the vet's help.

Once the mare has got rid of the afterbirth (has cleansed), her hindquarters should be cleaned with warm disinfectant. The vulva should be checked for tears. If the vulva is torn, it will need stitching by the vet. The mare should be given a bran mash with added limestone for extra calcium and then left in peace to enjoy her new foal.

15 The foal and young horse

The First Few Days

Foals born in the field are often brought in with their mothers so that they can be observed at three-hourly intervals for the first day so as to ensure that all is well. The mare will appreciate some cut grass while she is stabled. Providing the weather is fine, on the second day mare and foal can go out for exercise, and on some studs the foal may run loose behind the mare.

The foal will have to have a foal-slip (small head collar) put on at some point in the first week. The foal-slip is secured with an assistant holding one arm round the chest and the other arm round the quarters. If the foal is hard to catch, the mare can be so placed as to corner it. Right from the start, leading is safer than allowing the foal to run loose. For early leading, a soft web line is placed through the foal-slip and passed back to the left hand. At first, this hand steadies the foal by being placed across its chest. A second loop of similar material is dropped around the foal's quarters and is held on the loins. The mare should be led by an assistant and the foal kept by the mare's flank. Thus, the mare's leader is in front of the foal, its leader and the mare are on either side of it, and the back-strap is behind it.

After a week or two, if the mare and foal are coming in at night, the foal will become accustomed to being led and then one person will be able to handle both mare and foal. However, assistance may still be needed at the field gate. The fit of the foal-slip must be checked regularly: foals grow fast!

It is essential that the foal gets an adequate supply of milk. In spite of careful feeding, it may be that the mare does not produce enough. Alternatively, the mare may be tender or nervous and will not allow the foal to suck. Some mares need to be held, sometimes with a foreleg held up, while the foal is suckling. However, such

Fig. 15.1 The foal learns to be obedient and walk freely forward in its first week.

measures are rarely necessary, and even nervous or tender mares will generally come round within a few days. When suckling, the mare takes most of the hind weight on the leg next to the foal, so tilting her pelvis and making it easier for the foal to get to the teats.

Some mares are put back into work after about three weeks. It is then important that the foal should be allowed to suck every three hours. The normal pattern of sucking is a five- to ten-minute feed every two hours during the day, but less frequently at night. After taking nourishment, often the foal will relieve itself and then sleep for up to half an hour.

If either mare or foal should die or the mare have no milk, the National Foaling Bank should be consulted at once (telephone: Newport (Shropshire) 811234). This organisation has vast experience of artificial rearing and of fostering.

During the first hour of life, the foal will breathe faster than is normal, but should settle to a respiration rate of 20 to 30 per minute. This is about twice as fast as its mother's respiration rate.

The pulse will be about 80 at birth but go up to 140 as it struggles to find its feet. The pulse rate will settle to nearly 100 beats per minute for a day-old foal and will be down to under 50 for a yearling. The temperature should be 38.3 to 38.6°C (101.0–101.5°F).

Diseases of the Foal

Behavioural Disorders

Symptoms: Excessive nervousness, muscular twitching, shivering, nodding, staggering, lack of suck reflex, convulsions, incessant chewing, etc.
Causes: May be meningitis (inflamed skin surrounding the brain) or dummy syndrome (neonatal maladjustment syndrome, N.M.S.) caused by brain damage at birth.
Treatment: Call the vet. With good nursing under veterinary direction the foal may recover completely but it may well be lost.

Diarrhoea

Symptoms: Scouring.
Causes: A tummy upset typically caused by a chill or by a change in the mare; for example, when a mare comes into season on her foal-heat about nine days after foaling, the foal will often scour for a day or two.
Treatment: Simple treatments are available from the vet; in particular, the electrolytes (body salts) must be maintained.

Entropion

Symptoms: Ingrowing eyelid.
Cause: Congenital (born with it) imperfection.
Treatment: The vet can turn the eyelid outwards and secure it with a stitch until it settles correctly.

Haemolytic Disease

Symptoms: Sleepiness, yellow (jaundiced) or pale membranes, failure to suck, listlessness, red urine.
Cause: The red blood cells in the foal are destroyed by the antibodies in the colostrum from the mare.
Treatment: The foal will need an immediate blood transfusion. Once the mare's colostrum has finished (she must be milked out

regularly and the milk discarded), then the foal may use her milk again.

Hyperflexion or Weakness of Lower Limb

Symptoms: Pasterns and/or fetlocks hyperflexed (overbent) or weak (point of fetlock sinking so that the ergot nearly touches the ground).
Causes: Suspensory tendons and ligaments may be slightly long or short, or the muscle tension may be too weak or too strong.
Treatment: Mild cases will often right themselves. Severe cases may be aided by corrective boots carefully fitted and regularly changed. In areas of very high value foals such as Newmarket there are specialists who will provide such a service. Later on, tiny corrective shoes may be fitted with, for example, extended roll toes to sink the heel on to the ground for contracted tendon treatment.

Infectious White Scour

Symptoms: Scouring (diarrhoea) within a day or two of birth, covering the foal's buttocks in yellow/grey matter which smells unpleasantly.
Cause: A bacterial infection of the gut.
Treatment: Keep the hindquarters clean by regular washing in warm soapy water. A protective cream such as udder cream for dairy cows can be used to prevent scald. Maintain scrupulous hygiene. Burn the soiled bedding. Always treat this foal last in the round. The vet must be called and treatment given to control the infection and to cope with the dehydration which is inevitable with diarrhoea.

Joint Ill

Symptoms: Swelling and failure of the navel to dry up. Swollen joints and stiffness. Lack of appetite and symptoms of pain. The condition can rapidly deteriorate and cause death.
Cause: Blood poisoning by infection through navel.
Prevention: To avoid this disease or any failure of the colostrum to give early protection, it is common practice for the vet to examine a newborn foal and to give it an antibiotic injection. Also, foaling-box hygiene must be excellent and the foal's navel should be treated at birth.
Treatment: Call the vet at once.

Parrot Mouth and Cleft Palate

Symptoms: See page 9. The newborn foal must be examined for deformities such as these.
Cause: These are congenital abnormalities.
Treatment: As with any congenital abnormality, the vet must advise whether it will come right of its own accord, whether it may be operated on or treated, or whether the foal should be put down as not viable.

Pneumonia

Symptoms: Fast breathing, fever, coughing.
Cause: Inflamed lungs due to infection. Occurs more commonly in stuffy, poorly ventilated housing.
Treatment: Antibiotics, good nursing and warm but well-ventilated housing.

Sleepy Foal Disease

Symptoms: Fever, sleepiness, rapid respiration, failure to suck, loss of strength.
Cause: A specific bacterial infection.
Treatment: None possible. This disease is usually fatal.

Snotty Nose (Rhinopneumonitis)

Symptoms: Cold, catarrh, cough, nasal discharge.
Cause: A specific virus infection of the upper respiratory tract.
Treatment: Consult the vet.

Umbilical Hernia

Symptoms: At 4–6 weeks old a soft swelling appears at the navel.
Cause: The muscular ring, through which the vessels passed to form the umbilical cord, fails to close after birth and abdominal contents protrude.
Treatment: The vet will decide if and when treatment is necessary. The condition may right itself within twelve months. If the opening is squeezing the swelling (a strangulated hernia), then it will cause pain and need prompt surgical attention.

Worms

Symptoms: Loss of condition and failure to gain weight.
Cause: Worm parasites (as discussed in Chapter 8). Adult horses

may not show the effects of worms because they have a certain degree of immunity. Young stock are much more susceptible.

Treatment: (a) Rear young stock on pasture that is as clean as possible from parasite infection. (b) Keep the mare regularly treated so that she is not carrying a heavy worm burden. *Note*: some anthelminthics (anti-worm drugs) are not suitable for pregnant mares. (c) Maintain a regular monthly dosing programme from six weeks of age to cover both strongyle and ascarid worms, and bots in autumn.

Discipline

A rigid procedure is essential when several people are leading mares with foals or young stock. For example, if three mares with foals were led to a paddock, and the first handler, immediately on entering the paddock, were to release his mare and foal, the mare might well buck, kick, knock into the foal and then gallop off, thus causing problems for the other leaders, whose charges would want to join in the fun. All stud work involves care, thought, attention to detail and good discipline.

The early handling of foals is the beginning of their training and they must be taught good manners. It is important that this handling should be kindly but firm. Nothing is more confusing to animals than people who are inconsistent, or liable to be moody, or lose their tempers. A foal that rears or paws should be disciplined by slapping its chest and scolding it. A foal that bites (even in play) is best disciplined by pulling its whiskers rather than by hitting its muzzle. Violence and ill-temper must be avoided at all costs, as must over-indulgence.

When foals are two months old, they may be shown in hand with their mother. For showing it is important that the foal should be polite and lead well. The foal should be accustomed to being stood up in front of the mare and to having the mare stood up in front of it. When being shown, mare and foal will have to be separated to some degree, so that the judge can observe each move without the other masking the picture. All of this handling will help the foal to become accustomed to both humans and discipline. Foals must also learn to have their feet picked out and this will prepare them for having their feet trimmed by the farrier.

Weaning

Horses in the wild tend to live as family groups. As the colts mature, the mare will reject them, but a filly may remain with her mother for several years.

During the first winter after foaling, it is natural for the mare's milk flow to decrease markedly, and so the foal must be nutritionally independent. In stud management, it is usual to wean foals in September, when the grass loses much of its food value. However, if the mare is not again in foal, the foal need not be weaned at this stage. During weaning, the foal should be in good health and be showing independence.

In the month before weaning, the foal must get used to concentrates. Both mare and foal can be fed twice daily or, alternatively, 'a creep' can be introduced. A creep is a device for allowing the foal to feed, but not its mother. One type is a feed bowl attached to the fence, which has an adjustable narrow opening so that only the tiny muzzle of the foal can fit in to eat the food. A second type of creep is based on a narrow opening through which the youngster can pass but not the mother. However, for horses a wider opening is often used, but with an adjustable height bar across it so only the foals can pass under it. Where creeps are used, the foals must be attracted and introduced to the idea. This can be done by using highly palatable food to tempt the foal, which must also be introduced to the location of the feeder.

During weaning it is helpful if there are several mares and foals together in the field, which must be safely fenced. One of the mares is removed quietly one morning and she is placed with a companion in a far-distant well-fenced fairly bare field. Her foal is left with its friends and soon settles down. A day or so later, another mare or two may be taken. One mare should be left as long as possible as a guardian and disciplinarian.

An alternative method, which is traditional, but which tends to set back the foal due to the trauma, is to shut the foal in a stable. Probably the foal will try to climb out, and so the top door must be shut or else covered with a grille. There should be nothing in the stable that could catch up the foal, and so hay is fed on the floor. Water and concentrates are fed in safe containers that do not spill easily. When turning out, such foals may seek their mothers so they should have known company and a well-fenced field.

Colts should be separated from mares and fillies before the first spring after their birth because their natural tendency to mount the females is not to be encouraged. Colts may become fertile as two-year-olds but usually will be castrated before then.

Nutrition

When considering nutrition of young stock, the first essential is to decide on both short- and long-term aims and objectives. Is the foal to be a show winner as a foal, as a yearling, as a two-year-old, etc? Is the foal to be sold at weaning, or as a yearling, or when? Is the horse to be a competition winner? What sort of contest? All of these questions should have been considered before conception, but circumstances change and so reconsideration is necessary.

Two opposing facts must be balanced. First, the winner in the show ring or the most admired horse at a sale tends to be the most precocious animal – big, well advanced and possibly heavy-topped. To have a horse in this condition for a sale day is one thing, but to so maintain it for a show season has long-term dangers, particularly for one- to three-year-olds. Secondly, horses that have been overfed and have advanced too fast as youngsters tend not to make the best mature animals. On the other hand, it is a false economy not to feed the foal up to its growth potential. It is important that its nutrition allow it to reach full potential. The general long-term aim is good-quality food in plentiful supply so as to produce a big, strong, fit but not fat youngster.

Epiphysitis, which causes round and sometimes warm joints, is a danger to be avoided. The growing plates (epiphyses) on the long bones of the leg are just behind the bearing ends. These growing plates can get swollen by jarring. This is a particular danger with heavy-topped young stock on hard ground. The condition may also be caused by a lack of calcium, or a poor calcium : phosphorus ratio, or by bran inhibiting calcium uptake.

Young stock in winter getting a hay and cereal diet may lack some essential amino acids (e.g. lysine and methionine). To overcome this, a proprietary rearing mix may be used, or alternatively, a home mix which includes dried milk, dried grass or soya-bean meal. The home mix will also need mineral and vitamin supplements. The winter feeding should be so good that the foal moves smoothly on to spring grass on an equal plane of nutrition.

An old rule of thumb for Thoroughbred foals was to feed, daily, 1 lb (444 g) of concentrates for each month of age.

Old pasture rather than new is best for young stock. Ideally, the pasture should be free from ruts or deep winter tread marks which could strain the young foal's limbs. The sward can contain up to 10% clover; a higher ratio could interfere with bone metabolism. Lush spring grass (early bite) may be high in vitamin A, and as this depresses vitamin D uptake, rickets could result. Additional vitamin D is therefore useful in spring. Slightly acid pastures are better for good mineral uptake, and so over-liming must be avoided.

Education and Training

During the first three years of its life, the horse's bones are comparatively soft and are growing fast. Consequently, the horse should not be subjected to weight-carrying or strenuous work until

Fig. 15.2 A well-grown yearling moves pleasantly and freely forwards showing good in-hand training.

the third year. Abnormalities are likely to occur in the bones and joints if this precaution is not taken.

Flat-race horses are subjected to earlier training, but they carry minimal weights and work on straight or gently curving tracks. However, it should be noted that they have a very bad soundness record.

If a young horse is ruined through mismanagement, it will have to suffer for the remainder of its life through no fault of its own. Mismanagement is also costly to the owner because of all the costs that have been incurred in producing the youngster.

Probably, these costs will have been minimised by allowing the horses to live out, thus saving on labour, which inevitably reduces the amount of handling, unless they have been through a sale or have been shown. The quality of handling must be excellent even if it is minimal. A horse that has received strict but fair discipline during its most formative years will be easier to break and will therefore be set back less by the breaking process.

The procedure for breaking and initial training is outside the scope of this book. The authors' intention is to help readers to understand more about management, and the best management is *management with understanding*.

Appendix: The rationale of feeding horses

Rationing in six steps

Step one: *'How big is the horse or pony?'*
Estimate horse's bodyweight

Method   A   – Table of Weights (see Table A.1).
          B   – Tape Measure and Table.
          C   – Weigh-tape.
          D   – Weigh-bridge.
          E   – Guess.
Measure in kg (50 kg = 1 cwt)

Step two: *'How much can it eat each day?'*
Check horse's capacity
Method   Bodyweight in (kg)

$$\frac{\text{Bodyweight in (kg)}}{100} \times 2.5 = \text{capacity in kg}$$

Example A 16 hh, seven-year-old riding horse weighs 500 kg

$$Capacity = \frac{500}{100} \times 2.5 = 12.5 \text{ kg (28 lbs) of hay and concentrates per day}$$

Guesstimate: Horse's height in hands $\times$ 2 = capacity in lbs.

Step three: *'How much hay per day?'*
Provide energy for maintenance

$$Calculation \quad 18 \text{ MJ} + \frac{\text{Bodyweight (kg)}}{10} = \text{requirement}$$

Provide this from fresh grazed grass or conserved grass e.g. hay, haylage, silage. (See Table A.2.) However, don't feed too much hay, leave room for production.

Rough guide

	Hay (%)	Concentrates (%)
Maintenance	100	0
Light work	70	30
Medium work	50	50
Hard work	30	70
Fast work	25	75

Example Our Riding Horse.

$$\text{Energy for maintenance} = 18 + \frac{500}{10} = 68 \text{ MJ of digestible energy} \quad \text{(DE)}.$$

Table A.2 shows: average grass hay = 9 MJ of DE

$$\therefore \frac{68}{9} = 7.6 \text{ kg of hay per day (about 17 lb)}$$

= About 60% of capacity which is O.K. for light/medium work

Step four *'How much concentrate feed?'*
Provide energy for production

For *Work* per day, for each 50 kg bodyweight add MJ of DE:

Light work	+ 1 e.g. One hour walking.
	+ 2 e.g. Walking and trotting.
Medium work	+ 3 e.g. Some cantering.
	+ 4 e.g. Schooling, dressage and jumping.
Hard work	+ 5 e.g. Hunting 1 day/week.
	+ 6 e.g. Hunting 2 days/week.
Fast work	+ 7 e.g. 3-day eventing.
	+ 8 e.g. Racing.

For *Lactation* per day, for each 50 kg bodyweight add:

For first 3 months + $4\frac{1}{2}$ MJ of DE.
For next 3 months + $3\frac{1}{2}$ MJ of DE.

N.B. All diet changes must be gradual.

For *Pregnancy* per day, add:

+ 12% for the final $\frac{1}{3}$ of gestation or last 3 months.

For *Growth* per day, add:

Young stock over 1 year – feed at maintenance ration for their expected weight at maturity.

Up to 1 year provide 13 MJ of DE per kg of food and feed to capacity.

Example Our riding horse in light/medium work.

$$\text{Energy for production} = \frac{500 \text{ kg of bodyweight}}{50} \times 2 = 20$$

Table A.2 shows: barley = 17 MJ of DE

$$\therefore \quad \frac{20}{17} = 1\tfrac{1}{4} \text{ kg of barley per day for work (nearly 3 lbs)}$$

Step five *'Is the food suitable?'*
Provide sufficient protein

For *Maintenance* – 7.5 – 8.5% crude protein in ration

For *Production*

Light work	7.5 – 8.5%
Medium work	7.5 – 8.5%
Hard work	9.5 – 10%
Fast work	9.5 – 10%

For *Lactation*

First 3 months	14%
Next 3 months	12%

For *Pregnancy*

Final $\tfrac{1}{3}$ gestatation	11%

For *Growth* – For all growth check protein quality.

Suckling foal	16 – 18%
Weaned foal	14.5 – 16% (6 months +)
Yearling	12 – 14% (12 – 18 months)
One – Two	10 – 12% (18 – 24 months)
Two – Four	8.5 – 10% (24 – 48 months)

Example Our riding horse, seven years old:

Getting: 7.6 kg of hay + 1¼ kg of barley per day
Table A.2 shows the percentage of crude protein (CP) in the
feed so with the feed given above the %CP per day is:
Av. grass hay 7.6 kg × 8 = 61
        barley   1.25 kg × 11 = 14

8.85 into   75 = 8.5%CP

N.B. The nutritive values shown in Table A.2 provide a guide
based on averages, but individual food samples may show
variation.

Step six *'Mixing Art with Science'*
Check and adjust

1. Check that the ration contains sufficient roughage
Supply 0.6 kg roughage per 100 kg bodyweight or more. (This is
a minimum.)

Example Our riding horse weighing 500 kg.

$$\frac{500}{100} \times 0.6 = 3 \text{ kg of hay as a minimum}$$

2. Check the horse's condition
By eye and/or tape or weighbridge.
Is the horse gaining or losing weight?
Do you want it to do so?
What is its ideal performance weight?
Adjust the ration as necessary.

Example We would like our riding horse to have a bit more 'cover
over his ribs'. So we will increase the concentrates. One extra kg of
barley gives:
11% crude protein ⟶ Muscle
16 MJ of DE ⟶ Fat or energy

3. Check the horse's temperament
A Thoroughbred may need up to 10% extra food.
Native or draught stock may need 10% less.

N.B. Calm and routine – save food.

4. Check the environment
Feed extra under bad conditions both in field or in stable, e.g. cold stable, thin bedding, thin rug. A cold horse needs extra energy to keep warm.

5. Check for parasites
Feed only the horse. The commonest parasites are worms, especially round white worms in foals, red worms in all other stock.

6. Check the horse's efficiency as a converter of food
Horses reared on worm infested pastures may have permanently damaged guts and will always need extra care with their nutrition.

Horses with sharp teeth and sore mouths will not chew their food properly for good digestion.

7. Check that the horse has plenty of fresh water
Lack of water reduces capacity and digestion.

8. Check food quality and container hygiene
Mouldy hay, musty corn, tainted food, dirty manger; all depress appetite. Badly harvested, badly conserved, badly stored food loses quality, loses vitamins, loses appeal.

Food contaminated by vermin can carry disease.

9. Check that the horse is not lacking minerals or vitamins
Apart from salt, unless the horse is growing or under stress, it will not necessarily need supplements any more than humans do.

10. Check that the horse enjoys his food.

Height (hh)	Type	Approximate weight			Girth (cm)
		(kg)	(lbs)	(Other)	
10	Pony	200	450		135
13	Foal/Weaner	200	450	4 cwt	
12	Pony	300	700		150
13	Pony	350	800	7 cwt	160
14	Horse Yearling/ Pony	400	900		170
14.2	Pony	450	1,000		175
15	Hack	450	1,000		175
14.2	Cob	500	1,000	½ tonne	184
16	Thoroughbred	550	1,200		190
16	Hunter	600	1,350		196
16	Hunter	650	1,450		
16.2	Draught or Shire Horse	1,000	2,240	1 tonne	

Table A.1 Guidelines for estimating horse's bodyweight

	Crude protein (%)	Digestible energy (MJ/kg)
Cereals		
Oats	12	14
Barley	11	17
Maize	10	18
Protein		
Soya meal	50	15
Dried milk	36	18
Linseed	26	24
Field beans	26	16
Grass meal	12–18	13
Intermediate foods		
Wheatbran	17	12
Sugar Beet Pulp	10	15
Hays		
Good grass hay	10	11
Av. grass hay	8	9
Poor grass hay	4	8

Note: this is only a guide and a lot of variation is found in practice.

Table A.2 Nutritive value of some common foodstuffs.

Index